BEST BUYS
IN USED CARS

Third Edition

Jim Mateja

Bonus Books, Inc., Chicago

Dedicated to Sue, Brad, Gina and Dana

T*hanks to those who provided counsel, wisdom, and a statistic or picture or two.*

Dave Bowman, Allied Aftermarket Division of Allied Automotive [Fram]; R.J. "Tony" Mougey, Champion Spark Plug Co.; Ray Windecker, Ford Motor Co.; Bill Fair and John Perduyn, Goodyear Tire & Rubber Co.; Jerry Cizek, Chicago Automobile Trade Association; Jim Baumhart and Ed Przylucki, Better Business Bureau of Metropolitan Chicago; and dealers Ken Bennett, Kevin Briley, Dick Everman, Steve Foley, Joe Hennessy, Dick Hoskins, Jim Jennings, Lou Kole, Jim Lyng, Jim, Rick and Rob Mancuso and John Miller.

---

99   98   97   96   95   5   4   3   2   1

Library of Congress Cataloging-in-Publication Data: 95–081005

International Standard Book Number: 1–56625–049–8

Bonus Books, Inc.
160 East Illinois Street
Chicago, Illinois 60611

Printed in the United States of America

CONTENTS

INTRODUCTION

In the 16 years that I've been writing about cars and the auto industry, I've never had anyone come up to me and say: "I'm going out to buy a car this weekend and I'm really looking forward to it."

Rather, I'm typically told: "I'm going out to buy a car this weekend. My God, what do I do?"

The purpose of this book is to tell you what to do.

The fear that haunts us all when we go shopping for a car is that we'll make a big mistake—we'll either end up buying a double sour lemon tart or we'll pay too much for what is otherwise a good car. Most people going into a dealership are convinced that they are going to pay too much for the car and often they do (though that doesn't keep them from going home later and boasting to family and friends about what a great deal they just got). If you think used car prices are high, the breakdown of 1987 model prices at the end of this chapter should make you feel better.

What we all fear is being taken.

I've written this book to help you avoid being taken when you go shopping for a used car. I'll do it by telling you what to look for and what to look out for—and how to do both.

I'll begin by telling you why buying a used car can be a good idea and a wise investment, provided you heed the warning signs that can save you money in the long run.

I'll take you through the inspection procedure that you should follow and provide you with tips on how to spot the danger signs that are staring you in the face, such as the trailer hitch on the rear bumper that indicates that the brakes and transmission have seen their better days, especially if there are a dozen decals on the rear window boasting of trips to Yosemite and the Grand Tetons and you're looking at the car on a lot in Orlando, Florida.

I'll give you some personal preferences on good used cars you should consider buying and a long list of the lemon tarts — including make and model — that you should avoid at any price and why. And I'll tell you why you should stay away from all repos (repossessed cars) and rental cars.

I will describe some of the games that car dealers and car manufacturers play and how you can protect yourself against both.

I'll also tell you why you should never buy any car without taking it out for a test drive, preferably in the company of a trusted mechanic.

And because buying a car is only half the fun and caring for it is the other half, I'll provide you with some tips on service and maintenance, including how to "read" your sparkplugs for signs of major problems and how to "read" your tires for signs of future troubles.

I'll also include some information on a subject that you probably aren't even aware of — counterfeit parts. They are as phony as a $3 bill and worth as much, except that when you purchase and install a set of counterfeit brakes and they don't work, nobody is left laughing.

And since finding the right car depends on defining what kind of car you need in the first place, I will include a list of questions that you should ask and answer before you go out shopping for a car. I have put these at the back of the book as Appendix B.

What you're not going to find in this book is a nice easy formula to follow. I've heard from people who purport to have the entire car buying experience whittled down to A-B-C but that's a bunch of hokum. Any easy A-B-C approach is easy only because it leaves out the most important and least controllable element in car-buying — emotion.

Emotion is having psyched yourself up for the car buying experience and having rehearsed the familiar ploy, "Just looking," and then finding yourself blubbering like an idiot the first time you see that shiny red sports car and begging the salesman to let you buy it.

Emotion is telling ourselves to be rational and logical and then having all reason and logic disappear by falling victim to the first salesperson who greets us at the door with a friendly smile and an offer of a nice hot cup of coffee.

Emotion is being undone by ourselves and it happens to the best of us.

For instance . . . let me relate *my* last car buying episode.

My wife and I were invited to a party at an automobile museum. Within moments after arriving I should have realized something was up but I failed to react to the warning signs.

One was that my wife started bringing me champagne, not in the tiny little glasses the bubbly usually is sipped from, but in huge water mugs meant for gulping. And this was the woman who for 20 years had always waited until we approached the threshold of a party to announce: "And remember, you have to drive home, so lay off the hootch."

But this evening she was getting leg cramps from running off to the champagne fountain and I was getting tennis elbow from imbibing.

The next morning I woke up and through the haze found my wife all smiles — a clue that either something was wrong or I was in the wrong house.

"I was so nervous about the car I couldn't sleep," she giggled.

I looked out the window to see if I had pulled the car into the driveway or into the privet hedge again.

"What car are you talking about?" I asked.

"Why, the Corvette you bought me last night," she replied.

I dropped my head back on the pillow and tried to remember.

"Did I say why I bought you a 'Vette?" I finally asked.

"I said it matched the color of my eyes and you said: 'Buy it,' " she cooed.

Emotion is buying your beloved spouse of 20 years a 'Vette because it matches the color of her eyes when you can't see out of your own.

The other thing you're not going to find in this book is a list of ways to outwit the salesman. That's like trying to outcon a con man and it ain't gonna happen. You're playing his game on his grounds and he knows all the tricks and all the rules and you're not even certain of what the game is. For one thing, it includes that cup of coffee that you so innocently accepted when you stepped through the door. As you'll learn, booze isn't the only beverage that can be used to get you to buy a car.

What I want you to be able to do is to walk into a dealership and not beat *yourself*, to be able to keep a clear and rational head and know how to act in your own behalf and not in the best interest of the salesperson or the dealer.

If I can get you to do that, you will buy a good car.

Prices of 1987 model domestic automobiles
Base price excluding options, freight and taxes

Chevrolet

Mini: Chevette [RWD]
2-door hatchback	$ 4,995
4-door hatchback	$ 6,099

Subcompact: Cavalier [FWD]
2-door coupe	$ 7,255
4-door sedan	$ 7,449
4-door wagon	$ 7,615
2-door CS hatchback	$ 7,978
4-door CS sedan	$ 7,953
4-door CS wagon	$ 8,140
2-door RS convertible	$13,446
2-door RS coupe	$ 8,318
2-door RS hatchback	$ 8,520
4-door RS sedan	$ 8,499
4-door RS wagon	$ 8,677
2-door Z-24 coupe	$ 9,913
2-door Z-24 hatchback	$10,115

Subcompact: Nova [FWD]
4-door sedan	$ 8,258
5-door hatchback	$ 8,510

Compact: Beretta [FWD]
2-door coupe	$ 9,925

Compact: Corsica[FWD]
4-door sedan	$ 8,995

Compact specialty: Camaro [RWD]
2-door sport coupe	$ 9,995
2-door Z-28 sport coupe	$12,819

Intermediate [FWD]: Celebrity
2-door coupe	$ 9,995
4-door sedan	$10,265
4-door wagon	$10,425
4-door 3 seat wagon	$10,672

Intermediate [RWD]: Monte Carlo
2-door Monte Carlo LS coupe	$11,306
2-door Monte Carlo SS coupe	$13,463
2-door Monte Carlo SS aero coupe	$14,838

Standard size: Caprice [RWD]
4-door sedan	$10,995
2-door Classic coupe	$11,392
4-door Classic sedan	$11,560
4-door Classic Brougham sedan	$12,549
4-door Classic Brougham sedan L	$13,805
4-door Classic wagon	$11,995

Sports specialty: Corvette [RWD]
2-door	$27,999
2-door convertible	$33,172

Pontiac

Subcompact sports specialty: Fiero [RWD]
2-door 2-seater coupe	$ 8,299
2-door 2-seater Sport coupe	$ 9,989
2-door 2-seater SE coupe	$11,239
2-door 2-seater GT coupe	$13,489

Mini: T1000 [RWD]
2-door hatchback	$ 5,959
4-door hatchback	$ 6,099

Subcompact: Sunbird [FWD]
4-door sedan	$ 7,999
4-door wagon	$ 8,529
2-door SE coupe	$ 7,979
2-door SE hatchback	$ 8,499
2-door SE convertible	$13,799
2-door GT coupe	$10,299
4-door GT sedan	$10,349
2-door GT hatchback	$10.699
2-door GT convertible	$15,569

Compact: Grand Am [FWD]
2-door coupe	$ 9,299
4-door sedan	$ 9,499
2-door LE coupe	$ 9,999
4-door LE sedan	$10,199
2-door SE coupe	$12,659
4-door SE sedan	$12,899

Compact specialty: Firebird [RWD]

2-door coupe	$10,359
2-door Trans Am	$13,259

Intermediate [FWD]: 6000

2-door coupe	$ 9,995
4-door sedan	$10,499
4-door wagon	$10,899
2-door LE coupe	$11,099
4-door LE sedan	$12,389
4-door LE wagon	$11,499
4-door STE sedan	$18,099
4-door SE sedan	$12,389
4-door SE wagon	$13,049

Intermediate [RWD]: Grand Prix

2-door Grand Prix coupe	$11,069
2-door Grand Prix LE coupe	$11,799
2-door Grand Prix Brougham	$12,519

Full size [FWD]: Bonneville

4-door Bonneville sedan	$13,399
4-door Bonneville LE sedan	$14,866
4-door Bonneville SE sedan	$15,806
Safari wagon [RWD]	$13,959

Buick

Subcompact: Skyhawk [FWD]

2-door Custom hatchback	$ 8,965
2-door Custom coupe	$ 8,522
4-door Custom sedan	$ 8,559
4-door Custom wagon	$ 9,249
2-door Limited coupe	$ 9,445
4-door Limited sedan	$ 9,503
4-door Limited wagon	$ 9,841

Compact: Somerset [FWD]

2-door coupe	$ 9,957
2-door Limited coupe	$11,003
4-door Skylark sedan	$ 9,915
4-door Skylark Limited sedan	$11,003

Intermediate [FWD]: Century

2-door Custom coupe	$10,844
4-door Custom sedan	$10,989
4-door Custom wagon	$11,478
2-door Limited coupe	$11,397
4-door Limited sedan	$11,593
4-door Estate wagon	$11,998

Intermediate [RWD]: Regal

2-door coupe	$11,562
2-door Limited coupe	$12,303

Personal luxury: Riviera [FWD]

2-door coupe	$20,337

Standard size: LeSabre [FWD]

4-door sedan	$13,438
2-door Custom coupe	$13,616
4-door Custom sedan	$13,616
2-door Limited coupe	$14,918
4-door Limited sedan	$14,918
LeSabre Estate wagon [RWD]	$14,724

Standard size: Electra [FWD]

4-door	$16,902
2-door Park Avenue	$18,577
4-door Park Avenue	$18,769
4-door T-Type	$18,224
Electra Estate wagon [RWD]	$17,697

Oldsmobile

Subcompact: Firenza [FWD]

2-door coupe	$ 8,541
2-door S hatchback	$ 8,976
4-door sedan	$ 8,499
4-door wagon	$ 9,146
2-door GT hatchback	$11,034
2-door LX coupe	$ 9,639
4-door LX sedan	$ 9,407

Compact: Calais [FWD]

2-door coupe	$ 9,741
4-door sedan	$ 9,741
2-door Supreme	$10,397
4-door Supreme	$10,397

Intermediate [FWD]: Ciera

2-door S coupe	$10,940
2-door LS coupe	$11,747
4-door sedan	$10,940
4-door Brougham	$11,747

4-door wagon	$11,433
4-door Brougham wagon	$12,095

Intermediate [RWD]: Cutlass Supreme

2-door coupe	$11,539
4-door sedan	$11,539
2-door Brougham	$12,378
4-door Brougham	$12,378
2-door Salon coupe	$12,697

Personal luxury: Toronado [FWD]

2-door coupe	$19,938

Standard size: 88 [FWD]

2-door Royale coupe	$13,639
4-door Royale sedan	$13,639
2-door Brougham coupe	$14,536
4-door Brougham sedan	$14,536
4-door Custom Cruiser wagon [RWD]	$14,420

Standard size: 98 [FWD]

4-door Regency sedan	$17,371
2-door Regency Brougham coupe	$18,388
4-door Regency Brougham sedan	$18,388

Cadillac

Coupe deVille [FWD]	$21,316
Sedan deVille [FWD]	$21,659
Eldorado coupe	$23,740
Fleetwood Brougham sedan [RWD]	$22,637
Fleetwood limo	$36,510
Fleetwood Formal limo	$38,580
Fleetwood d'Elegance	$26,104
Fleetwood Sixty Special	$34,850
Seville	$26,326
Cimarron	$15,032
Allante	$54,700

Chrysler/Plymouth

Subcompact: Horizon [FWD]

4-door America	$ 5,799

Subcompact: Sundance [FWD]

2-door coupe	$ 7,599
4-door sedan	$ 7,799

Subcompact sports specialty: Turismo [FWD]

2-door hatchback	$ 7,199

Compact: LeBaron GTS [FWD]

4-door hatchback	$10,152
4-door hatchback Premium	$11,767

Compact: Reliant [FWD]

2-door sedan	$ 7,879
4-door sedan	$ 7,879
2-door LE sedan	$ 8,364
4-door LE sedan	$ 8,364
4-door LE wagon	$ 8,808

Intermediate [FWD]: LeBaron

2-door coupe	$11,295
2-door convertible	NA
4-door sedan	$10,707
Town & Country wagon	$12,255

Intermediate [FWD]:

4-door Caravelle	$ 9,813
4-door Caravelle SE	$10,527

Intermediate [RWD]: Gran Fury/New Yorker

4-door Gran Fury Salon	$10,598
4-door Fifth Ave.	$15,666

Full size [FWD]: Chrysler

4-door New Yorker	$14,396

Dodge

Subcompact: Omni [FWD]

4-door America	$ 5,799

Subcompact: Shadow [FWD]

2-door coupe	$ 7,499
4-door sedan	$ 7,699

Subcompact sports specialty: Charger [FWD]

2-door hatchback	$ 7,199
2-door Shelby hatchback	$ 9,840

Subcompact sports specialty: Daytona [FWD]

2-door hatchback	$ 9,799
2-door hatchback Turbo Z	NA

Compact: Lancer [FWD]

4-door hatchback	$ 9,852
4-door hatchback ES	$10,806

Compact: Aries [FWD]

2-door sedan	$ 7,879
4-door sedan	$ 7,879
2-door LE sedan	$ 8,364
4-door LE sedan	$ 8,364
4-door LE wagon	$ 8,808

Intermediate [RWD]: Diplomat

4-door Diplomat Salon	$10,598
4-door Diplomat SE	$11,678

Full size [FWD]: Dodge 600

4-door sedan	$10,010
4-door SE sedan	$10,672

Ford

Subcompact: Escort [FWD]

2-door Pony	$ 6,586
2-door GL	$ 6,949
4-door GL	$ 7,163
4-door GL wagon	$ 7,444
2-door GT	$ 8,815

Subcompact sports specialty: EXP [FWD]

2-door 2-seater coupe	$ 7,745
2-door 2-seater sport coupe	$ 8,920

Subcompact sports specialty: Mustang [RWD]

2-door LX sedan	$ 8,271
2-door LX hatchback	$ 8,690
2-door LX convertible	$13,117
2-door GT	$12,106
2-door GT convertible	$15,852

Compact: Tempo [FWD]

2-door GL	$ 8,160
4-door GL	$ 8,310
2-door GL sport	$ 8,909
4-door GL sport	$ 9,059
2-door LX	$ 9,321
4-door LX	$ 9,520
2-door 4WD	$10,044
4-door 4WD	$10,194

Intermediate specialty: Thunderbird [RWD]

2-door coupe	$13,028
2-door sport	$15,065

2-door LX	$15,357
2-door turbo coupe	$16,600

Intermediate: Taurus [FWD]

4-door sedan L	$10,650
4-door wagon L	$11,870
4-door GL sedan	$11,622
4-door GL wagon	$12,802
4-door MT5 sedan	$12,074
4-door MT5 wagon	$12,654
4-door LX sedan	$14,633
4-door LX wagon	$15,243

Standard size: LTD Crown Victoria [RWD]

2-door coupe	$14,709
4-door sedan	$14,349
2-door LX coupe	$15,378
4-door LX sedan	$15,410
4-door Crown Victoria wagon	$14,315
4-door Country Squire wagon	$14,567
4-door LX Crown Vic wagon	$15,489
4-door LX Ctry Squire wagon	$15,471

Mercury

Subcompact: Lynx [FWD]

2-door L	$ 6,716
2-door GS	$ 7,094
4-door GS	$ 7,308
4-door GS wagon	$ 7,590
2-door XR3	$ 8,897

Compact: Topaz [FWD]

2-door GS	$ 8,664
4-door GS	$ 8,814
2-door GS sport	$ 9,319
4-door GS sport	$ 9,469
4-door LS sedan	$10,266

Intermediate: Cougar [RWD]

2-door LS	$13,360
2-door XR-7	$15,660

Intermediate: Sable [FWD]

4-door GS sedan	$12,340
4-door GS wagon	$12,904
4-door LS sedan	$14,544

4-door LS wagon	$15,089
Standard size: Grand Marquis [RWD]	
4-door GS	$15,163
2-door LS	$15,478
4-door LS	$15,621
4-door GS wagon	$15,500
4-door LS wagon	$16,029
Luxury: Lincoln [RWD]	
4-door Town Car	$22,837
2-door Mark VII	$23,246
2-door Mark VII LSC	$24,837
4-door Continental	$25,484
2-door Merkur XR4Ti	$17,832

American Motors

Subcompact: Alliance [FWD]	
2-door	$ 6,399
4-door	$ 6,599

3-door hatchback	$ 6,399
2-door L	$ 6,925
4-door L	$ 7,200
2-door L convertible	$11,099
3-door hatchback	$ 6,975
5-door hatchback	$ 7,250
2-door DL	$ 7,625
4-door DL	$ 7,900
2-door DL convertible	$12,099
3-door hatchback DL	$ 7,675
5-door hatchback DL	$ 7,950
3-door GS hatchback	$ 8,499
2-door GTA sedan	$ 8,999
2-door GTA convertible	$12,899
Compact utility: Eagle [4WD]	
4-door sedan	$11,485
4-door wagon	$12,301
4-door Limited wagon	$13,033

Prices of 1987 model Japanese automobiles
Base price excluding options, freight and taxes—Oct. 1, 1986

Toyota

Tercel Liftback 4-speed	$ 5,898
Tercel Liftback Deluxe auto	$ 7,788
Tercel Liftback Deluxe 5-speed	$ 7,358
Tercel 5-dr. Liftback Deluxe auto	$ 7,988
Tercel 5-dr. Deluxe wagon auto	$ 8,768
Tercel 5-dr. Deluxe wagon 5-speed	$ 8,398
Tercel 4WD wagon automatic	$ 9,938
Tercel 4WD wagon 6-speed	$ 9,588
Tercel 4WD wagon SR5 6-speed	$10,638
Mr 2 2-seater 5-speed	$12,548
Mr 2 2-seater automatic	$13,238
Mr 2 2-seater 5-speed T-Top	$13,738
Mr 2 2-seater automatic T-top	$14,428
Corolla 4-door Deluxe sedan 5-sp	$ 8,178
Corolla 4-door Deluxe sedan auto	$ 8,608
Corolla 4-door LE sedan 5-speed	$ 9,278
Corolla 4-door LE sedan automatic	$ 9,878
Corolla 5-door Deluxe Liftback auto	$ 9,038
Corolla SR5 coupe 5-speed	$ 9,548
Corolla SR5 coupe automatic	$10,148
Corolla GT-S coupe 5-speed	$10,368
Corolla FX-16 Liftback 5-speed	$ 9,678
Corolla FX-16 Liftback automatic	$10,368
Corolla FX-16 Liftback GTS 5-spd	$10,668
Corolla FX-16 Liftback GTS auto	$11,358
Celica ST 5-speed	$10,598
Celica ST automatic	$11,198
Celica GT coupe 5-speed	$12,038
Celica GT coupe automatic	$12,638
Celica GT Liftback 5-speed	$12,288
Celica GT Liftback automatic	$12,888
Celica GT convertible 5-speed	$16,798

Celica GT convertible automatic	$17,398
Celica GT-S coupe 5-speed	$13,978
Celica GT-S coupe automatic	$14,668
Celica GT-S Liftback 5-speed	$14,328
Supra 5-speed	$19,990
Supra automatic	$20,680
Supra Sport automatic	$20,990
Supra Turbo 5-speed	$22,260
Supra Turbo automatic	$22,950
Supra Sport Turbo 5-speed	$23,260
Camry 4-door Deluxe sedan 5-spd	$10,798
Camry 4-door Deluxe sedan auto	$11,478
Camry 4-door LE sedan automatic	$13,398
Camry 4-door wagon 5-speed	$11,488
Camry 4-door wagon automatic	$12,168
Camry 4-door wagon LE automatic	$14,168
Cressida 4-door 5-speed	$19,350
Cressida 4-door automatic	$19,350
Cressida 5-door wagon automatic	$19,410

Nissan

Sentra 2-door	$ 6,199
Sentra 2-door E	$ 7,499
Sentra 2-door XE	$ 8,399
Sentra 2-door hatchback E	$ 7,349
Sentra 2-door hatchback XE	$ 8,449
Sentra 4-door E	$ 7,899
Sentra 4-door XE	$ 8,749
Sentra 4-door GXE	$ 9,699
Sentra wagon XE	$ 9,099
Sentra wagon GXE	$10,049
Sentra wagon XE 4WD	$ 9,999
Sentra XE coupe	$ 9,399
Sentra SE coupe	$10,399
Pulsar XE	$10,599
Pulsar XE automatic	$11,059
Pulsar SE	$11,799
Stanza 4-door E	$10,499
Stanza 4-door GXE	$11,899
Stanza 4-door hatchback XE	$11,499
Stanza wagon XE	$11,349

Stanza 4WD wagon XE	$12,749
200-SX notchback XE	$10,849
200-SX hatchback XE	$11,199
200-SX hatchback SE	$14,499
Maxima SE sedan	$16,099
Maxima GXE sedan	$16,099
Maxima GXE wagon	$16,999
Maxima SE sedan auto seat belts	$16,499
Maxima GXE sedan auto seat belts	$16,499
Maxima GXE wagon auto seat belts	$16,999
300-ZX 2-seater	$18,499
300-ZX turbo coupe	$21,399

Honda

Civic hatchback standard	$ 5,799
Civic hatchback DX	$ 7,489
Civic hatchback DX automatic	$ 7,989
Civic hatchback Si	$ 8,899
Civic 4-door sedan 5-speed	$ 8,455
Civic 4-door sedan automatic	$ 9,180
Civic 4-door wagon	$ 8,330
Civic 4-door wagon automatic	$ 8,830
Civic 4-door 4WD wagon	$ 9,695
Civic CRX HF two seat coupe	$ 7,639
Civic CRX two seat coupe	$ 7,975
Civic CRX two seat coupe automatic	$ 8,475
Civic CRX Si	$ 9,395
Accord hatchback DX 5-speed	$ 9,795
Accord hatchback DX automatic	$10,325
Accord hatchback LXi 5-speed	$12,785
Accord hatchback LXi automatic	$13,315
Accord 4-door sedan DX 5-speed	$10,625
Accord 4-door sedan DX automatic	$11,155
Accord 4-door sedan LX 5-speed	$12,799
Accord 4-door sedan LX automatic	$13,329
Accord 4-door sedan LXi 5-speed	$14,429
Accord 4-door sedan LXi automatic	$14,959
Prelude 5-speed	$11,995
Prelude automatic	$12,525
Prelude Si 5-speed	$14,945
Prelude Si automatic	$15,475

Mazda

323 hatchback	$ 6,099
323 hatchback DX	$ 7,699
323 sedan DX	$ 8,299
323 sedan LX	$ 8,899
323 wagon DX	$ 8,899
626 4-door sedan DX	$ 9,899
626 4-door sedan LX	$11,699
626 4-door sedan GT	$13,149
626 2-door coupe DX	$ 9,949
626 2-door coupe LX	$11,899
626 2-door coupe GT	$13,449
626 4-door touring sedan LX	$12,399
626 4-door touring sedan GT	$14,049
RX-7	$14,199
RX-7 LX	$15,799
RX-7 Sport	$15,749
RX-7 GXL	$18,449
RX-7 2 plus 2	$14,699
RX-7 2 plus 2 LX	$16,299
RX-7 2 plus 2 GXL	$18,949
RX-7 turbo	$20,399

Subaru

Standard hatchback 4-speed	$ 5,398
GL hatchback 5-speed	$ 7,588
GL hatchback automatic	$ 8,093
GL hatchback 4WD	$ 8,293
DL sedan 5-speed	$ 8,808
DL sedan automatic	$ 9,518
GL sedan 5-speed	$ 9,838
GL sedan automatic	$10,643
GL 5-speed turbo	$12,383
GL-10 sedan automatic	$12,348
GL-10 sedan 5-speed turbo	$13,628
GL-10 sedan automatic turbo	$14,053
GL sedan 4WD	$10,308
GL sedan 4WD automatic	$11,138
RX sedan 4WD turbo	$13,833
GL-10 sedan 4WD turbo	$15,068
GL-10 sedan 4WD turbo automatic	$15,243
DL wagon 5-speed	$ 9,208
DL wagon automatic	$ 9,918
GL wagon 5-speed	$10,238
GL wagon automatic	$11,043
GL-10 wagon 5-speed turbo	$12,783
GL-10 wagon automatic	$12,748
GL-10 wagon 5-speed turbo	$14,028
GL-10 wagon automatic turbo	$14,453
DL wagon 4WD	$ 9,598
GL wagon 4WD 5-speed	$10,708
GL wagon 4WD automatic	$11,538
GL wagon 4WD 5-speed turbo	$13,388
GL-10 wagon 4WD turbo 5-speed	$14,688
GL-10 wagon 4WD turbo automatic	$15,643
DL coupe 5-speed	$ 9,108
GL coupe 5-speed	$10,138
GL coupe automatic	$10,673
GL coupe turbo 5-speed	$12,593
GL coupe 4WD turbo 5-speed	$10,608
GL coupe 4WD automatic	$11,143
GL coupe Full time 4WD turbo 5-speed	$14,053
XT coupe DL 5-speed	$ 9,593
XT coupe GL 5-speed	$11,518
XT coupe GL automatic	$11,943
XT coupe GL-10 turbo 5-speed	$14,573
XT coupe GL-10 turbo automatic	$14,998
XT coupe GL-10 4WD turbo 5-sp	$15,648
XT coupe GL-10 4WD turbo auto.	$16,098
BRAT GL 4WD	$ 8,338

WHY USED?

A matter of record

So your old clunker has lost its clunk. That makes it time to buy a new car. Is your savings account in worse shape than your clunker? Then it's time to consider buying a used car.

You've probably heard it said that the reason a person buys a used car instead of a new one is to save money. That is the main reason but it is not the only one.

You've probably also heard people insist that they wouldn't buy a used car because they don't want to buy "somebody else's troubles." Well, some of the smartest car buyers I've ever encountered deliberately buy only used cars. Why?

Because used cars have a track record. Those who sit back for a year or two and then buy used aren't as likely to "buy someone else's troubles" as they are to "learn from someone else's troubles."

Now that doesn't mean that you can tell from simply eyeballing it whether a 1982 Olds 98 will last 10 years or 10 minutes. But it does mean that when you see the little name badge on the decklid that says "Diesel," you have sufficient warning that you could be getting yourself into trouble. If you don't think so, ask the man or woman who has tried to get rid of one over the last couple of years.

Buying used can not only save you money, it can save you heartache and grief if you pay attention to the market and keep track of whatever troubles are reported. Because a car that proves to be undesirable when new doesn't get any better when it is used.

As examples of what I mean by track record, let's take a look at some cars that were unveiled with great promise yet turned out to be real duds.

In the fall of 1970, Chevrolet introduced the subcompact Vega as a 1971 model. It was to be GM's import fighter against the likes of the Volkswagen Beetle from West Germany. People still weren't sure what a Toyota Corolla really was and Nissan was still called Datsun. Honda sold motorcycles and a little box on wheels some people laughingly referred to as a car.

Vega was in trouble right from the start. James Roche, then chairman of GM, couldn't even pronounce the car's name right. He called it "Veega."

People rushed out to get the Vega and 255,929 of them were sold the first year. By the 1973 model year Chevrolet rang up sales of more than 440,000 units. But trouble was catching up with it. The car featured an aluminum block engine, a deliberate lightweight design so Chevy could boast about the mileage. But the engine was warping, owners found, unless a plastic reservoir tank was added under the hood to trap coolant overflow. So the tanks were added.

Then owners found, perhaps from all the moisture of the overflow, that the front fenders were prone to rust. You've surely heard Lee Iacocca's famous slogan for Chrysler: "Buy a car, get a check." With Vega, it was: "Buy a car, get a set of fenders, quick."

Hundreds of thousands of people who bought the early models tried desperately to dump them. I owned a 1972 Vega wagon. The plastic reservoir tank prevented any engine warping and boil-over problems. But I became an expert in replacing fenders. If you ever want to determine if a person once owned a Vega do a Sherlock Holmes and look at his hands. If you see scars along the backs of the knuckles you can tell he changed a fender or two.

Vega wasn't alone. Everyone raved when Honda brought out the subcompact Accord in 1975. At the time it was the epitome of what a small car should be — roomy, comfortable, high on mileage, long on looks, a pleasure to drive. But it wasn't until November 1981 that the 1975–1979 Honda Accord fender rust problem was finally addressed by the Japanese automaker.

And talk about track records. GM converted its 350 cubic inch V-8 gas engine to diesel power and offered it in Oldsmobile cars in the 1978 model year. Wow! Thirty miles a gallon in an Olds 98.

That engine was one of the most influential engines GM ever offered. Within a matter of years it served to bring together people from all walks of life, races, religions and creeds in a joint crusade — to sue GM because the thing wouldn't stay together. Name the diesel engine part and eventually it wouldn't work. Dealers in many cases stopped taking them in trade. Today most used car value books leave a blank under the heading of what one is worth.

In the 1980 model year, GM thought it had come up with the perfect small, high mileage line of compact cars, the so-called X-bodies, the Chevrolet Citation, Buick Skylark, Olds Omega and Pontiac Phoenix. Consumers obviously felt the same way because they bought the cars in such numbers that GM actually ran out of such necessities as replacement oil filters for buyers who simply wanted to change their oil.

It took a few years but then allegations began to surface that the front-wheel drive cars' brakes had a tendency to lock in a panic stop. The X-body series lost favor with the public but it attracted the attention of the National Highway Traffic Safety Administration, which went to court in an attempt to force GM to recall those cars.

And let's not let Ford Motor Company off the hook. Its subcompact Pinto came out in the 1971 model year as competition to the Vega. It, too, sold in great numbers, until news reports began circulating that in a rear end collision the position of the car's fuel tank next to the rear axle turned the Pinto into a potential bomb.

In most of these cases, it took a while for the public to learn of the cars' problems. One reason is that the public wasn't as critical of cars back when a 1971 Vega or Pinto could be purchased for about $2200.

Now, with new car prices hovering around $12,000, the motoring public has come to expect perfection—and if it doesn't get it, it expects the automakers to right their wrongs at once. The public is being more critical and consumer groups are being more vocal and Detroit, Japan and Europe are being more quickly called to account when they make a mistake.

But that's just one reason. There's also another one.

Look again at some of those problem cars mentioned above: the 1971 Chevy Vega, the first year for that car on the market; the 1971 Ford Pinto, the first year for that car on the market; the 1980 GM X-body cars, the first year for those cars on the market.

See the trend?

You worry about buying a used car and getting someone else's problems. I worry about buying a *new* car and getting Detroit's troubles. I worry about buying the *new* car the first year on the market and serving as Detroit's guinea pig.

In the spring of 1981 General Motors brought out its new subcompact J-body Chevrolet Cavalier as an '82 model. Like the X-cars before it, Cavalier was a hit—until buyers realized they had a problem. The car had been designed three years earlier when gas prices started rising and GM expected motorists to be paying $2 to $3 a gallon for fuel by 1982. As a result, GM put a 1.8 liter four cylinder carbureted engine under the Cavalier hood. The Cavalier was rated rather high in mileage because the 1.8 liter four was indeed stingy with a gallon of gas. But the reason it didn't burn much fuel was that the engine was so underpowered that the car couldn't crawl. Forget about not having the pep to make it up hills. The Cavalier had trouble making it up inclines.

It was so severely underpowered that owners began to scream at its sluggishness. Gas cost only $1 to $1.25 a gallon and so consumers were interested in miles per hour rather than miles per gallon.

GM solved the problem by offering a beefed up 2 liter, fuel injected version of the 1.8 liter but not until the 1983 model year. The buyers of those 1982 models were stuck with them.

What's the Detroit mentality about consumers serving as guinea pigs? I asked an engineer at GM what the 1982 J-body Cavalier car owners were to do about the dramatic problem of underpowered engines. His response was: "They should buy an '83." And he was serious.

If you are bold and adventuresome, if your life-long wish has been to tempt the unknown, to be a test pilot or to serve as an astronaut, then maybe being an unpaid guinea pig for Detroit is something you should look into. Personally, if I went into a grocery store and the lady passing out samples stuck the tray in my face and said, "Here, take one and see if *you* die, but first hand me over $10,000," I'd have some second thoughts about what I was getting into.

Yet there are consumers who buy a brand new car without any track record and say they've done so because the alternative of buying used is no better than "buying someone else's troubles."

The other thing you might consider doing is buying Japanese. The Japanese are very considerate of the American consumer and so they design, develop and build a car and then sell it in Japan for one or two years to work out all the bugs before they rename the car and begin marketing it in the United States. That way their fellow countrymen suffer so that we don't. Nice touch, that.

So why don't American automakers reciprocate and sell

their cars in Japan for a year or two before peddling them here? With taxes and duties on U.S. cars in Japan, a $14,000 Dodge Caravan was selling for $30,000 when I visited Tokyo in 1985. You can't test what they can't buy.

The National Automobile Dealers Association estimates that the price of a new car in this country is about $12,000 out the door. Hertz Corporation calculates the average price of a used car off the lot at $5431. (See summaries at the end of this chapter.) Waiting a bit to see how a car performs not only saves you the trouble of fixing or repairing a mechanical ailment, it can put $6500 in your pocket.

Cheap? Nope. Smart? You betcha.

There's another reason why I consider used car buying preferable in many cases to the purchase of a new car. Actually, there are a number of reasons and they include: Grand Am, Calais, Fiero, MR-2, Caravan, Voyager, Aries, Reliant, Taurus, Sable, Tempo, Topaz, Galant, Starion, Sentra, Stanza, Cavalier Z24 and a host of others. Each of these cars was introduced after 1980.

Beginning with the 1978 model year, the American auto industry has been engaged in a massive downsizing program that still isn't over. In addition to making its cars smaller and lighter in order to make them more fuel efficient to meet federal mileage laws, the industry has added such features as front-wheel drive, four-wheel drive, anti-lock brakes and fuel injected engines — as well as producing more mini vans.

Imports, already leading the way in fuel economy because the cars already were mostly small and light, have turned to the same technological innovations as the domestics, as well as multi-valve 16 and 24 valve engines. Not only have the names of the cars changed dramatically in the last few years, the cars have undergone nearly total transformation.

There's a story they tell about Lee Iacocca and Henry Ford II that fits in here because it makes a point. Ford and Iacocca have hardly been friends while here on this earth.

The day comes and Ford dies and goes to heaven. At the Pearly Gates he meets St. Peter and asks him for assurance that Iacocca isn't there. St. Peter assures him Iacocca isn't.

As Ford enters the gates, a Chrysler New Yorker breezes by, a huge machine with a gent in the back seat smoking a giant cigar. Ford rushes back to St. Peter and says: "I thought you assured me that Iacocca isn't here. I just saw a New Yorker go by with a guy in the back seat smoking a huge cigar."

To which St. Peter replied: "Don't worry, that was God. He just thinks he's Iacocca."

The point is that Iacocca isn't God. No car manufacturer is. And the best proof of that is that God created his product in six days and then stopped. The car manufacturers are still mucking around with theirs and not always for the better. Of course, considering the shape the world is in, maybe God ought to consider a little retooling, too.

To put downsizing in perspective, the 1975 Chevrolet Nova was considered a compact size car. It was built on a 111 inch wheelbase, which means that the hub to hub measure was 111 inches. The 1986 Olds 98 is built on a 110 inch wheelbase, or one inch shorter than the '75 Nova. Yet the 1986 Olds 98 is called a full size car.

So the family that owned an 1984 Olds 98 built on a 116 inch wheelbase, tipping the scales at more than 5000 pounds and powered by a 350 cubic inch V-8, now finds the same car built on a 110 inch wheelbase, slimmed down to about 3500 pounds and a 3.8 liter (cubic inches no longer are spoken) V-6 rests under the hood.

At GM today, the Pontiac Grand Am is considered a compact size car. It's built on a 103 inch wheelbase. How times have changed.

Downsizing has created as much confusion as the new names, new engines and new transmissions. For example, while the 103 inch wheelbase Grand Am is a compact at GM, the 99 inch wheelbase Tempo is a compact at Ford and the 100 inch wheelbase Dodge Aries is a compact at Chrysler.

The 103 inch wheelbase Plymouth Caravelle is considered midsize at Chrysler but at Ford the 106 inch wheelbase Taurus is called a full size model. See what I mean about confusion? That's why lots of people are sitting back or buying used cars they are familiar with until the automakers get it straight among themselves. Good idea.

The Hertz Corporation does an annual survey of the used car market. In looking over the 1985 survey, I spotted a few things that I would like to share with you. One is that most used cars — 37 percent — were purchased at new car dealerships that sell their trade-ins. But dealers are declining as the source. In 1985, 5.89 million used cars were purchased from new car dealers, representing a decline from the 7.92 million purchased from them in 1980. At the same time, 29 percent of the cars, or 4.62 million, were purchased from friends and relatives, up from 2.68 million in 1980. The rest — 9.3 million — were bought from used car lots and private parties and strangers, an increase of 26 percent from 1980. That has significance as I'll explain later.

A couple other key figures are that 10 percent of those who bought a used car in 1985 never bought a used car before. That's a lot of people, more than 1.59 million of them. And the percentage of women buyers has risen dramatically, from 20 percent to 30 percent. And of the used cars purchased, 79 percent were domestics, down from 83 percent the prior year, meaning that just as with new cars, consumers are buying more imports.

Hertz found that 74 percent of all used car buyers say that the reason they go that route is that the used car is "less expensive." And Hertz has the figures to back them up.

According to Hertz, the average price of a used car purchased in 1985 was $5431 and while that was up $25 from the average purchase price of $5406 a year earlier, where do you find a new car at that price unless you swallow the bullet and go for a new $3990 Yugo—if you can swallow it.

Hertz calculated that the new car that was purchased for $9834 would cost the used car buyer one year later $7734—a 21 percent savings. That same car purchased as a two year old car would cost $5768—a 41 percent savings. And if the same car was purchased as a three year old used car, the cost would be $4351 or a 56 percent savings over the cost when the car was new.

Wait a minute, you say. The new car has a warranty and you can get lots of things fixed free. That used car is older and will require service and maintenance at my own expense.

Sure you can expect to pay for upkeep but keep in mind that the first owner absorbed all the depreciation and hefty interest charges on the new car loan. Taxes on a $9834 car at six percent are $590. Taxes on a $4351 three year old car are $261, or $329 less. You can line your den with oil filters for $329 and that's just with the savings on taxes.

New Car Calendar Sales/Selling Price Summary

Year	Cars sold per dealer	Average selling price
1980	327	$ 7,530
1981	326	$ 8,850
1982	316	$ 9,910
1983	373	$10,725
1984	420	$11,100
1985	444	$11,925
1986	456	$12,950
1987	406	$13,450
1988	421	$14,100
1989	392	$15,400
1990	379	$15,900
1991	342	$16,050
1992	353	$17,100
1993	373	$18,200
1994	394	$19,200

Source: National Automobile Dealers Association.

Used Car Calendar Year Sales/Selling Price Summary

Year	Sales	Avg. price
1980	9.76 million	$ 3,798
1981	9.91 million	$ 4,226
1982	9.78 million	$ 4,829
1983	10.63 million	$ 5,248
1984	12.41 million	$ 5,693
1985	13.36 million	$ 5,814
1986	13.54 million	$ 5,978
1987	13.26 million	$ 6,485
1988	14.60 million	$ 7,131
1989	14.61 million	$ 7,430
1990	14.18 million	$ 7,543
1991	14.27 million	$ 7,861
1992	15.14 million	$ 8,314
1993	16.30 million	$ 9,131
1994	17.98 million	$10,140

Source: National Automobile Dealers Association.

AVAILABILITY AND PRICING

Detroit builds sport coupes and family sedans. Detroit builds convertibles, vans and mini trucks. Detroit builds four cylinder and six cylinder and eight-cylinder engines. Detroit builds cars that get 30 miles to the gallon and some that seem to get 30 gallons to the mile.

But Detroit *doesn't* build used cars. Neither does Japan, Germany, England or Italy, although you might get an argument on England and Italy from former Triumph or Fiat owners.

Used cars are manufactured by those who bought new cars and then three, four or five years later decided to trade them in on another new machine. Used cars are the machines that once were new. Just how used they are depends on the treatment given them. And just how many used cars are available in any given year depends on how many people are willing to part with them for a new machine.

So if you are in the market for a used car hope that new cars are selling well. Because the used car market suffers when new car buyers stay out of the market. When they aren't buying new, they aren't trading in potential used cars and the supply of used cars diminishes, with the usual results — prices go up.

The other factor related to new cars that you should keep your eye on is prices. The prices of new cars impact the value of used cars in two ways. The first, obviously, is that the higher the new car price goes, the higher the subsequent used car price will rise. This may force the person such as yourself, who was thinking of buying a three year old car, to have to look instead at a four year old car.

The second effect is that the higher the price on the new car becomes, the more attractive — and valuable — the used car becomes as an alternative. Clearly, as new car prices continue to climb, the number of potential buyers forced into the used car market increases.

As for pricing, the practices of Detroit in recent years are worth a review. There isn't much you can do about them but I think you ought to at least know what's been going on. You might call it upsizing the downsized — or less car for more money.

It was in 1980 that domestic auto makers began changing their pricing policies. It used to be that customers could expect a big price increase each fall and that was it until the next model year.

The problem was that in 1980 the increase averaged more than $1000 per model and that had a number of unhappy effects. It upset the consumers who wanted to buy a car and it upset the dealers who wanted to sell cars. And it made Detroit look bad.

So the auto makers went back to the drawing boards and restructured their pricing methods. Rather than one whopping increase in the fall, they developed what you might call sequential pricing whereby they increased prices modestly in the fall and then two or three more times before the end of the model year. They also reduced the dealer discounts which are actually the dealer markups. So if the dealer discount was 22 percent on a full size car, it was now trimmed to 18 percent.

By doing this, the auto makers were able to still raise prices dramatically but put the brunt of it on both the consumer and the dealer rather than just on the consumers as in

Model Year Domestic Base* Price Summary

	1979	1980	1981	1982
GM	$5,868	$7,137	$8,430	$9,412
Ford	$5,705	$6,646	$7,515	$8,304
Chrysler	$5,372	$6,134	$7,260	$7,816
AMC	$4,750	$5,913	$7,093	$7,419
Industry avg.	$5,666	$6,732	$7,856	$8,694

	1983	1984	1985	1986
GM	$9,679	$10,306	$11,029	$11,551
Ford	$8,819	$9,672	$9,980	$11,109
Chrysler	$8,580	$9,118	$9,358	$9,839
AMC	$7,474	$7,498	$8,119	$8,261
Industry avg.	$9,157	$9,800	$10,278	$10,932

Source: Industry pricing figures

*Base prices exclude taxes, options and freight charges. Taxes and freight alone will add about $1,000 to the price, options another $1,000 to $2,000 or more.

the past. So if an auto maker wanted to raise its new car price by $1000, it would raise the price of the car to the consumer by $500 and reduce the dealer discount by three percent or $500. The two actions combined provided the auto maker with the $1000 it wanted.

But Detroit still wasn't satisfied. In 1980, the government imposed a price increase ceiling of 5.5 percent on all new cars which meant that that was as much as Detroit could raise its prices. So just before the end of the 1979 model year,

Detroit boosted prices so that the 5.5 percent would be applied to a higher base and the auto makers would get more money.

These pricing methods still hold. Detroit likes to boast each fall that it *only* raised prices by X percent over the previous model year. But here's how they're able to do that. Just before the end of the model year—in the last month, in fact—the auto maker raises prices by $500. The $10,000 Spritzel now lists at $10,500. One month later the Spritzel for the next model year comes out and the price is $11,000.

So now the Spritzel has an $11,000 price tag compared with the $10,000 of only a month ago. How much did the car go up? Only $500, says Detroit, because it compares year *end* prices with those at the *outset* of the next model year.

But that's not all. The auto maker says all-season tires have been made standard on the $11,000 Spritzel. Those tires had been a $200 option for the customer (perhaps they cost the factory $50). So now when calculating the price increase on the $11,000 Spritzel, Detroit adds the $200 cost of those tires to the $10,500 the car had sold for at the end of the prior model year and comes up with a $10,700 value. So Detroit says the new Spritzel was increased in price *only* $300 for the new model year ($11,000 minus $10,700).

The $10,000 Spritzel has gone up 10 percent to $11,000 in one model year. But Detroit, by its calculations, says the $10,700 Spritzel has gone up only 2.8 percent and announces that it has "held the line on prices." See how easy it is to curb inflation?

But wait . . . there's more. What if Detroit raised the price of the $10,000 Spritzel to $10,500 at the end of the year and then at the outset of the new model year priced the Spritzel at $11,000 but *deleted* the all-season tires that had been standard equipment on the car at both the $10,000 and the $10,500 levels and made them a $200 option?

Here's what happens: Detroit takes its $11,000 price tag and subtracts the year end $10,500 price and says it raised the price by $500. But of course, in effect, since you now have to pay $200 for those tires, the increase is actually $700. But who's to know? You? Now you do. And maybe you can use it as bargaining leverage.

More. What if Detroit raises the price of the Spritzel from $10,000 to $10,500 at year end and then at the outset of the new model year boosts the sticker price to $10,700. But in doing so, it adds the previously optional $200 all-season tires as standard equipment. Since the $10,500 car would have cost $10,700 with the optional tires and the new car is priced at $10,700, Detroit can now say that for the new model year it didn't raise prices at all despite the fact that the $10,000 Spritzel of a month ago now costs $10,700.

And one more "what if." What if the $10,000 Spritzel was raised to $10,500 at year end but stayed at $10,500 at the beginning of the new model year despite the addition of all-season tires as standard? Since the tires were worth $200 at retail and the $10,500 car didn't *go up* by $200, there must have been a price *reduction* and Detroit proudly announces that it reduced the price of its car by $200.

Simple, isn't it? And you thought the only thing that came off of those drawing boards was technological innovations.

Let's move on to used car availability . . . and some more bad news.

As we noted earlier, it stands to reason that the more cars that are sold new, the more that eventually will find their way onto the used car market. And, of course, conversely, the fewer sold new the fewer will be available as used cars a few years down the road.

I hate to have to be the one to tell you this but we're reaching one of those points in the road. Take a look at the summary of calendar year sales at the end of this chapter. Look closely at the domestic sales figures for each year from 1980 to 1985. The 1980–1982 domestic total is slim according to historic standards, especially in 1982. (Remember the Reagan Recession?)

What this means for the potential used car buyer is that those very low totals are about to catch up with the market. When? Now and for the next couple of years. Industry estimates are that most buyers are holding on to their new cars for four to six years before trading them in. So the '82s are coming and so are the '81s and '83s which aren't much better in terms of total numbers. It all adds up to slim pickings for the next few years — and consequent higher prices.

And don't look to imports for relief. Japan doesn't make a car called Rolaids. Not yet, anyway.

Look again at the New Car Sales Summary. Notice how the sales of imports have risen steadily from 2.3 million in 1979 to 2.8 million in 1985 while domestic sales have dropped from 8.3 million to 8.2 million. Impressive, isn't it? Also deceptive. Despite that increase and despite all the hootin' and hollerin' from Detroit over what imports are doing to the market, the fact is that number of imports sold represents just about one-fourth the number of domestic cars sold. So if the traditional import buyer who shops for used models were to be joined by the traditional domestic buyer who is now switching over to an import, the supplies wouldn't be able to handle the demand.

And don't look for great deals among imports because you will always have a couple of factors working against you. One is that Japanese imports, which account for about 80 percent of all foreign car sales in the U.S., have been under voluntary quota restrictions since 1980. These have kept prices artificially high. Another is the reaction of the Japanese themselves to the quotas. They have changed marketing strategies. Rather than focus on small, low-priced, high mileage economy cars, they've been promoting — and selling — luxury cars and sports models, cars that carry higher profit margins and help offset the limited volume of sales permitted under the quotas.

And then there's high finance — the impact that the value of the yen vs. the value of the dollar has on prices. Starting in late 1985 and running through calendar 1986, the value of the Japanese yen in relation to the U.S. dollar rose dramatically. The result was that the prices of most Japanese cars were raised four to five times by an average of $1400 just to offset the shifting values of the two currencies.

So what to do? Well, one thing, obviously, is to hang onto the old clunker and try to stretch its life out for another year or two. For one thing you can save money on insurance. There's no point in covering an old model car with low book value for collision, fire and theft. Save the money and put it aside to use as a down payment when you do get a car in a year or two.

And keep shopping around. Despite the odds, you may just find the right car at the right price, perhaps from a private seller who either doesn't know what the market situation is or has to get rid of the car for some personal reason.

After all, there is no fixed price on a used car like there is on a new one. The price is what the market will bear, which means what the salesman will tolerate and what you are willing to pay and where those two meet.

Just make certain that the car is the right car and that it is what it is represented to be. How to do that is what we will talk about in subsequent chapters.

New Car Sales Summary

Year	Domestics	Imports	Total
1970	7.1 million	1.3 million	8.4 million
1971	8.7 million	1.5 million	10.2 million
1972	9.3 million	1.6 million	10.9 million
1973	9.7 million*	1.7 million	11.4 million***
1974	7.4 million	1.4 million	8.8 million
1975	7.0 million	1.6 million	8.6 million
1976	8.6 million	1.5 million	10.1 million
1977	9.1 million	2.0 million	11.1 million
1978	9.3 million	1.9 million	11.2 million
1979	8.3 million	2.3 million	10.6 million
1980	6.5 million	2.4 million	8.9 million
1981	6.2 million	2.3 million	8.5 million
1982	5.7 million	2.2 million	7.9 million
1983	6.79 million	2.37 million	9.16 million
1984	7.95 million	2.42 million	10.37 million
1985	8.20 million	2.82 million**	11.02 million

Source: Industry sales statistics

*record high domestic total sales

**record high import total sales

***record high industry total sales

[record low domestic total sales 5.56 million in 1961]

[record import total sales share 27.8 percent in 1982]

[record low industry total sales 5.95 million—5.56 million domestics, 390,000 imports—in 1961]

COUPE OR CONVERTIBLE?

When you go looking for an apartment or a house, you generally do some homework in advance and establish a set of guidelines to follow to make certain that it will meet your needs. You decide, for example, that the apartment must have three bedrooms or the house must be within a short commute to the train.

Once you've found something within those parameters, you have a series of questions you want answered: Does the rent include utilities? Does the landlord allow pets? Is the house heated with gas or oil? When was the roof last repaired?

You should follow the same procedure with car buying. Before you ever leave the house to go off to the showroom, lot or the house of the private party with a car for sale, sit down — preferably with pen and paper — and figure out what kind of car you need and how much you are prepared, and can afford, to spend for it.

Actually, there is a series of questions you should ask yourself — and either answer or insist on the answers to — at every step of the car buying process. I've worked these up and included them in the last chapter of the book. Take a look at them. They will serve you as a guide and as a checklist for any car you buy, now or in the future.

In this chapter, we will focus on the questions you should ask and the factors you should consider to determine what kind of car you need. Basically, that comes down to who is going to use the car and what are they going to use it for. That may sound simple but it actually opens up a range of possibilities and a host of factors that you need to consider.

Let's start with the family car. This could be switched among several drivers and used for a variety of purposes. It could be used for running errands, for carting the kids — and the neighbors' kids — around, for car pools in all kinds of weather, for picking up the groceries on the weekly shopping trip, and for the family vacation.

In that case what you'll want to focus on is roominess, safety, respectable looks, ease of upkeep and maintenance and reliability. You may want to consider front-wheel drive if not four-wheel drive for the added traction on wet or snow-covered roads. Or if you've been thinking of a wagon, you might want to consider moving up to a van. For all their trendiness, the mini vans are great. They look, and for the most part act, better than wagons.

Obviously, one thing you don't want is a convertible. They may look great and they may seem great but they have all the practibility of a surfboard in the desert. You can't ride around with the top down in the winter and you shouldn't do it in the summer, unless you go in for fried brains. About the only time you can enjoy one in most parts of the country is in the fall and even then you have to keep the heater going full blast after dusk.

Frankly, I wouldn't recommend one for anyone. They're unsafe for kids and they're unsafe for teenagers. In fact, in the event of a rollover, they're unsafe for everybody. And they're not what I would like to leave sitting around in a shopping mall parking lot. If you're into convertibles, I have no special advice to give you other than what I've just said. Just follow the advice in this book and shop for the best one you can find for the price you're willing to pay. And then — Good Luck!

Besides the all-purpose family car, there are a number of special situations for which you might want a second car, such as running back and forth to the train station for the daily commute or to a construction job. In both cases, what you want primarily is reliability.

Forget about the mileage. What difference does it make if it gets only 15 miles per gallon instead of 25 if you're only using it for 10 to 25 miles a day? And forget about the looks. How good is it going to look anyway after a year of sitting out in the train station parking lot or at the construction site?

Concentrate instead on making certain that the engine and transmission are in good shape and that the car starts readily. What you want is that the car will start at 20 degrees below zero — not that it will get 20 miles to the gallon.

If you are a student — or the parent of a student — heading off to college, there's a different set of criteria to apply. Here your chief concern should be size for safety, load carrying ability and ease of maintenance.

What you don't want is a little subcompact. Sure, it gets 20 miles per gallon and the school is a thousand miles from home. But that's a case of fuelish being foolish.

Consider the scenario. The small car will be loaded to the brim. This will affect both the weight distribution and the visibility. You probably will be traveling alone since, with all your own stuff, there won't be any room for anyone else. And you'll probably be traveling tired since that's the normal condition of most students.

In that condition — and with your car in its condition — you are going to be out on the highway for hours, going head to hubcap with all those Knight-mares of the road, the screaming semis that go barrelling by you faster than a speeding bullet.

There's a school of thought out there that says "get young drivers a small car because not only will it save on mileage but a small car is easier to maneuver out of a potentially tight spot and you can avoid an accident that way."

I say — No Way. I know. I've been there.

I have driven hundreds of different cars over the past 16 years and I can tell you that when a semi-trailer comes up at you from behind on the expressway doing about 25 mph over the speed limit and you're sitting low behind the wheel of a Ford Escort with automatic transmission, you will suddenly wish that you were sitting in a Ford Crown Victoria with a 302 V-8. Try to out-maneuver a semi in a small car. If you survive, the first thing you will want to do is go back to the person who told you about small car maneuverability and maneuver his or her neck — with your shaking hands.

Higher mileage and lower gas costs in a small car I will grant you — up to a point. You will save about $200 or so a year in gas costs with a subcompact. And if you are a single adult, responsible only to and for yourself, and you spend most of your time tooling around the city and your primary concerns are a tight budget and tighter parking spaces, you might want to consider a subcompact. But keep in mind that all the things stated above about safety in size apply to you as well and with more and more maniacs running red lights — and if you run red lights you *are* a maniac — that is something that you ought to give some considerable thought to. Which is more important to you — your money or your life? Jack Benny had trouble making that choice. Should you?

In any case, if you are a college student with many miles to go before you sleep, skip the subcompacts and go for size and safety. And go domestic. Domestics are more forgiving if the oil isn't changed on a regular basis. And they are much easier to find parts for.

If the school is in a remote area, there aren't likely to be a wide range of import dealers available to help service and repair the car or independent mechanics capable of working on it.

With a domestic, if it's an Olds and it breaks down and the Olds dealer doesn't have the part, he can call the Pontiac or Chevy or Buick shop in town and get it. It'll be a different part number but it will be the same part.

But if a Toyota breaks down you will have to wait for the part to come from a Toyota distribution center at some remote point. So goodbye, weekend. And when the part does come you'll probably have to pay more.

There is one other category of driver that has to be discussed but that one is so special that it requires — and deserves — a chapter of its own. And that is what is coming next.

TEENAGERS AND CARS

Son Brad started getting restless about six months prior to his 16th birthday. The day was soon approaching and here he was without a car. It was unthinkable. No wheels and in just 24 weeks he'd have his driver's license.

If you soon will face the car buying experience with a teenage son or daughter read on — and carefully.

To a teenager, the primary consideration in buying a car is external looks and appearance. For most of them, if it looks like an economy model they won't touch it. The worst thing you can say to a teen when looking at cars in a lot is: "But that car is so practical." The only time a teen will listen to the old practicality speech is when he or she is in need of a quick infusion of capital. When pocket or purse is full again, practicality is forgotten.

What they love are the Firebirds, Camaros, Mustangs and Shelby Chargers. Or on the import side, the Celicas, Supras and RX-7s. And the old Triumphs and Fiats — mainly because nobody will have one like it. That's right. You won't find many Triumphs or Fiats on the road. In fact, you'll have to conduct a search party to find a mechanic to fix it or a store that has the part that just broke. Buy one of those rare old used cars and you'll find you're saddled with a car that spends more time in the mechanic's shop than the mechanic who owns it.

In Brad's case, I was smart enough for one of the few times in my life to use reverse psychology. Rather than tell him what to buy, I would let him talk himself out of the dogs and end up with a smart car on his own.

The father of a classmate ran a used car lot and that's where we headed. It was one of those garish facilities where most of the money is invested in multi-colored plastic banners swinging from rickety phone poles. The brilliant spotlight was fixed on the sheet of plywood that advertised the name of the lot. The softer lights, rather dim bulbs, were aimed at the cars. The lot wasn't paved. Instead, there was a thick layer of sand below each car, all the better to rake each day and cover up the oil and transmission fluid leaking onto it.

From about two blocks away, Brad spotted an orange Mustang. I was really surprised. The car looked super. The orange shined brightly. The kid may have picked a winner. The closer we came to the car, the more his eyes bulged and the more I started to feel ill. The car was a brilliant orange all right, except that the partially opened door was yellow along the lower door panel. And the sealant that had been used to put in the new windshield to replace the one obviously smashed in an accident was applied with a trowel and actually had drooled two inches down the glass.

I pointed out the orange over yellow paint job and the shoddy windshield replacement. Brad's response was: "Yeah, but what do think of the car?"

Reluctantly, he gave in — for a week.

Next up was a 1976 Firebird. And not just a Firebird but the Formula model complete with 440 cubic inch V-8. A friend owned it and was trying to sell it. Brad brought it home. It was a marvel. The friend had purchased it from a fellow who did a total body restoration. No plastic filler. All metal and the job was done beautifully. The car was left for Brad to drive a few days.

Brad asked Dad to take a test drive. The first thing I

noticed was that at acceleration the body swung wildly to the left. Same thing when braking. The tires told why: four different brands, sizes, and three different constructions — two radials, one bias belted and one bias ply.

The odometer told another story — 94,000 miles.

The price was right. The car was purchased. A new set of four matching tires was put on. The lurching went away. The car was fine — except for the fact that the 440 V-8 teamed with automatic was delivering only about 8 miles per gallon. After a few months of filling the tank on his own, Brad was ready for a different car.

My experience with Brad is similar to those all parents of teenagers will or have gone through. The numbers can be worn off the odometer or, as in Brad's case with the Firebird, approaching the second time around. The tires can be bald and the fuel economy measured in gallons per mile but if it looks good and carries one of the popular names mentioned above, the teen will promise to do homework or take out the garbage until the day he or she is married in order to get that machine. The promises last about as long as the tires, brakes and transmission but for a few days at least parents and teen are talking.

While you're talking, use what influence or leverage you have to push safety. Too often, because the teenager is still in school and earning minimum wage or not working at all, he or she will end up with an old clunker held together with baling wire. Brakes are poor, transmission is suspect and engine is overly fatigued. As a parent, you've spent 16 to 18 years of your life bringing the kid up, walking the floor when he or she was sick and then you allow them to go out in a car that should be in the junk yard. For shame!

Or you fall for the argument that young drivers should have a small car because they're easier to maneuver. What I said about that in the last chapter for college students goes double for teenagers. In spades. If you didn't read that chapter, go back and do so. And if you did read it and you're going to let your kid go out and buy a subcompact, read it again.

We're talking here about someone who has had little time behind the wheel, maybe a year or two of running back and forth to McDonald's or the movies in the family car. And now you're going to let them go out on the road with the 80,000 pound semis with just 2000 pounds of subcompact wrapped around them? For double shame.

Use whatever influence you have in trying to keep them out of the subcompacts or the convertibles or the clunkers. It won't be easy, it almost never is. Youth after all is the age of immortality. Young people never believe that anything bad is going to happen to them. It takes a while for them to understand just how many of their fellow immortals are looking up at the world from six feet under. Or out at it from a wheelchair. Permanently.

Anyway, now at least you know something about how teenagers think and feel about cars. And if you're a teenager who's been sneaking a look at what parents should know about you, now you know. (You won't know how they feel about you until you are a parent yourself.) Now try to get together somewhere in the middle. And if you succeed, let *me* know. Because then I'm going to sit down and write myself another book about how to get along with teenagers. And make a mint.

CARS TO LOOK FOR
PART 1

Sales leaders

Now that you know in general terms what kind of vehicle you need, the next step in the process becomes going out and finding it. Are there particular types or models of cars that you should look to buy . . . and other types you should never buy?

The answer is yes . . . on both counts. And that is what we are going to get into now.

A caller once asked me the value of an old car. I don't recall the name so let's call it a 1956 Spritzel. I didn't have the slightest idea what the car was worth but being a '56 I thought it must have some worth. To verify my guess, I called Greg Grams, the owner along with brother Bill of an antique automobile museum in Volo, Illinois, and members of the Antique Automobile Club of America.

Greg quoted me a value, which I remember as being rather low.

"But it's a '56," I said.

"Yes, but it was a dog in 1956 and when it comes to long term value of a car, a rule of thumb is that if it's a dog when it's new, it doesn't get a pedigree when it's old."

That's why, when shopping for a used car, your rule of thumb should be to focus on the winners, those cars that were the sales leaders when they were new.

Now being the top seller in the market in any given year doesn't automatically qualify a car as a best buy. The 1980

Citation proves that. Ditto the 1981 Chevrolet Chevette, a cramped, sluggish, lifeless looking car. Being the lowest priced car for a weekend in the Hertz or Avis fleet should tell you something.

But if you look closely at the Top Ten sales leaders over the years (see sales leaders tables at the end of this chapter), you will find, in nearly every case, reliable cars.

And there's something else to keep in mind about the sales leaders: they attract the attention of parts and components outfits. The more cars sold new, the more readily available you'll find repair and replacement parts for several years because those stores know they'll do a good business from used car owners. You can still go to most parts stores and get an oil filter for a 1975 Vega, or a headlamp for the '79 Ford Fairmont or a taillight for a 1978 Chevy Malibu.

That's also an argument for buying domestic rather than import, particularly if you live in the Midwest. The folks in California are able to drop into almost any auto store, discount house or mass merchandiser and pick up parts for their imported car. That's because half the cars sold in California each year are imports, particularly Japanese (except in Hollywood and Beverly Hills where they prefer European), and so it pays for those who handle parts to stock items for them.

The same is true on the East Coast although the preference is for European rather than Japanese.

But in the Midwest, it's a different story. Mid-Americans buy American. An import is treated like an invader. Few parts are stocked for them and most garages, service stations and independent mechanics are familiar only with domestic cars and will only work on them. The independent mechanic who is able and willing to work on an import is a specialist who works only on imports and he is often hard to find.

Domestics have another advantage and that is the commonality of parts. Under the names, and under the hoods, there's a lot of family heritage.

Take the 1985 calendar year sales leaders, for example.

The midsize Chevy Celebrity was the third best seller in the market with the Olds Ciera number four and the Buick Century number eight. All three are GM A-body cars with the same base engine and therefore the same basic parts. Need an engine component and you've tripled your chances of finding it. Won't the Chevy part differ from the Buick? Sure, Chevy will call it C-ABC and Buick will call it B-ABC.

Try finding a filter for any one year old German, Italian or French import. Try, for that matter, to find a filter for the majority of Japanese cars. Unless you live on the West Coast where more than half the cars sold are Japanese brands, many parts outlets or discount stores with automotive departments don't stock a large variety of foreign repair or replacement parts.

As a group, Japanese cars account for nearly 30 percent of total car sales in the U.S., but the individual makes account for relatively small numbers. The Toyota Corolla, Honda Accord and Nissan Sentra are the exceptions. Enough are being sold, as evidenced by their standings in the Top Ten, to justify parts stores carrying an ample inventory of replacement items.

Yes, sure you can find the parts for any car if you just go back to the dealer. If you have no qualms about paying $6 to $9 for an oil filter at the Toyota or Audi dealership versus $2.99 (on sale, which is often) for the 1975 Vega oil filter at a discount store, then you've won the argument. Personally, I'd rather pay $2.99 and use the savings to treat the family to a soft drink so I can sneak away for a beer.

Another factor most people overlook who support the "why worry, I'll get the part at the dealership" theory is that most dealership parts departments close early and often aren't open on weekends — just when *you* have the time or the need to buy a replacement part. Will you have to take off work an hour to buy an Audi or Volvo oil filter when the shop is open or would you like the convenience of running to your Sears, Ward, Penney, K-Mart, Venture or whatever store at 9 p.m. on a weeknight or early Sunday morning and find the part you want and get the job done now?

Not only do you assure yourself an adequate supply of such things as service and maintenance items when you buy one of the best selling cars, you also insure that there will be an ample supply of body parts still lingering around in the future.

So when the left front fender of the six year old Chevy Caprice is dented beyond recognition, you'll be able to find a replacement unit nearly anywhere. Big deal? Well, when you *can't* find a fender you have to bang out the old one and apply 100 pounds or so of plastic body filler as a less attractive and more time consuming alternative.

Also, not only will retail or discount stores or dealers have a ready access to those parts of the more popular models, it stands to reason that if worse comes to worst, the local junkyard will be able to come up with the part as well.

There is, however, another point worth noting about new car popularity and its effect on replacement parts. When a large number of any particular model is sold, it benefits eventual used car buyers. But it can, for a short period of time, play havoc with the new car owner.

Example: In mid-1979, General Motors Corporation introduced its X-cars, the compact, front-wheel drive Chevy Citation, Buick Skylark, Olds Omega and Pontiac Phoenix. The cars took the market by storm and GM couldn't build them fast enough to fill demand.

About six months after the cars were in the consumers' hands, the first problem of sales popularity raised its ugly head. Owners found that they could not locate oil filters for the simple chore of changing the fluid. The cars were selling so fast, the filters were all going into the new cars and even then GM was having a tough time meeting the demands of the assembly line.

In many cases, the oil change simply had to be delayed because there was a shortage of filters.

Not important? Only an oil filter?

You establish the life of the engine in the first 500 miles of driving. The first thing I would do after purchasing a new car is change the oil and filter.

So while looking at the sales statistics is a benefit in choosing a used car, it also can serve as a warning signal to the prospective new car buyer.

A more recent example is the Ford Taurus and Mercury Sable from Ford Motor Company. Those cars first appeared in December of 1985 as 1986 models. They represented the first new front-wheel drive full size cars at Ford. They also were another exercise in the aerodynamic styling look at Ford and were quickly accepted by the public.

In the first quarter of 1986, Ford's production of those two cars already was sold out through the second quarter of the year. Hot! But then the calls started coming in. A fuel pump went bad, the buyer said. Problems are to be expected, I replied. But, the buyer said, the car has sat at the dealership for three weeks and still no pump. Based on the demand, I had to say, that was to be expected.

Whenever a model is in big demand but low supply, the newspapers and magazines and trade books all report on the phenomenon. You should take note of that and back off awhile. Of course, I recommend that you lay off most cars their first year on the market, anyway.

Take a closer look at the car sales tables. You will note that with few exceptions a car seldom makes it into the Top Ten the first year it appears on the market. Ironically, one of the exceptions was the Chevrolet X-body Citation in 1979 although that was helped along by the fact that GM also counted in the Nova that the Citation was replacing.

There are two reasons why a new car seldom breaks into the sales elite immediately. One is that production start-up typically is slow. It takes about 18 months for a plant to reach full production capacity. The other is that consumers know better. They realize they should give a car at least a year on the road before jumping into it. They realize little is to be gained by serving as a guinea pig for the manufacturer.

If it's wise to stay away from a new car the first year it is on the market, then it's just as wise to stay away from the used car that represents the first year that model was on the market.

But didn't the manufacturer fix the mistakes? Didn't GM, for example, bring out higher power 1.8 liter and 2 liter four cylinder fuel injected engines in the subcompact J-body cars to replace the totally underpowered 1.8 liter carbureted engines they offered?

Sure they did—in 1983—the *second* year the cars were on the market. The 1982 J-cars didn't have the power to get out of their own way. And still don't.

Calendar Year Sales Leaders

1994

1. Ford Taurus
2. Honda Accord
3. Ford Escort
4. Toyota Camry
5. Pontiac Grand Am
6. Honda Civic
7. Saturn
8. Toyota Corolla
9. Chevrolet Cavalier
10. Nissan Sentra

1993

1. Ford Taurus
2. Honda Accord
3. Toyota Camry
4. Chevrolet Cavalier
5. Ford Escort
6. Honda Civic
7. Chevrolet Lumina
8. Ford Tempo
9. Pontiac Grand Am
10. Toyota Corolla

1992

1. Ford Taurus
2. Honda Accord
3. Toyota Camry
4. Ford Escort
5. Chevrolet Lumina
6. Chevrolet Cavalier
7. Pontiac Grand Am
8. Ford Tempo
9. Honda Civic
10. Toyota Corolla

1991

1. Honda Accord
2. Ford Taurus
3. Toyota Camry
4. Chevrolet Cavalier
5. Ford Escort
6. Chevrolet Lumina
7. Honda Civic
8. Toyota Corolla
9. Ford Tempo
10. Chevrolet Corsica

Calendar Year Sales Leaders—2

1990

1. Honda Accord
2. Ford Taurus
3. Chevrolet Cavalier
4. Ford Escort
5. Toyota Camry
6. Toyota Corolla
7. Honda Civic
8. Chevrolet Lumina
9. Ford Tempo
10. Pontiac Grand Am

1989

1. Honda Accord
2. Ford Taurus
3. Ford Escort
4. Chevrolet Cavalier
5. Toyota Camry
6. Ford Tempo
7. Nissan Sentra
8. Pontiac Grand Am
9. Toyota Corolla
10. Honda Civic

1988

1. Ford Escort
2. Ford Taurus
3. Honda Accord
4. Chevrolet Cavalier
5. Ford Tempo
6. Hyundai Excel
7. Chevrolet Celebrity
8. Nissan Sentra
9. Olds Ciera
10. Pontiac Grand Am

1987

1. Ford Escort
2. Ford Taurus
3. Honda Accord
4. Chevrolet Cavalier
5. Chevrolet Celebrity
6. Hyundai Excel
7. Olds Ciera
8. Nissan Sentra
9. Ford Tempo
10. Pontiac Grand Am

Calendar Year Sales Leaders—3

1986	1985
1. Chevrolet Celebrity	1. Chevrolet Cavalier
2. Ford Escort	2. Ford Escort
3. Chevrolet Cavalier	3. Chevrolet Celebrity
4. Olds Ciera	4. Olds Ciera
5. Honda Accord	5. Ford Tempo
6. Ford Tempo	6. Honda Accord
7. Ford Taurus	7. Chevy Impala/Caprice
8. Olds 88	8. Buick Century
9. Buick Century	9. Nissan Sentra
10. Chevy Impala/Caprice	10. Olds Cutlass Supreme

Source: Industry sales statistics.

CARS TO LOOK OUT FOR
PART 1

In school, the three R's that could get you into trouble were reading, 'riting and 'rithmetic. In used car buying, they are repos, rentals and recalls. There are those who say that one way to save a great deal of money is to shop for a rental car that most outfits now put on the market after a few months of use. Others look for repossessed cars through banks, savings and loan institutions, credit unions or any financial institution stuck with a car after the owner failed to make sufficient payments.

To both of these, I have one simple response:

No!

You won't find two better examples of abuse than rentals and repos. In six months' time a typical rental will undergo three years of abuse. In one week's time the car may have been subjected to three different drivers, each with unique driving habits and each caring less about caring for the car.

When I am driving my own car and a railroad crossing suddenly looms ahead, I slow down and gingerly approach the inevitable bump. I tiptoe over the obstruction and then proceed. If I'm driving a rental and the crossing is ahead, I don't even slow down.

Most rental outfits that do put their cars up for sale do so after a specified time or mileage. And most do their best to recondition the car and make it presentable for sale. Most of the ones they sell are the better ones from the fleet and usually are offered with some type of warranty. But, folks, they haven't just been used, they've been *used*.

Sometimes rental cars are sold through dealer auctions, which are weekly affairs held in various cities throughout the country where dealers go to peddle their trade-ins and purchase fresh stock for their used car lots. These are private sales open only to dealers and their representatives.

If a dealer's supply of used cars is low and the price of the rentals going through is low enough, he may stock up on some. That is why it is always wise to check on who the former owner of the car was and why you should *insist* on seeing the title listing the former owner before signing on the bottom line.

Be especially alert with the GM X-body cars. The Chevy Citation proved to be highly successful in sales to fleet customers and a large number of them were used that way. Of the 150,000 Chevettes sold in 1985, two-thirds of them were sold to fleets. So beware.

Even worse than the rental is the repo. These cars are taken back by the financing institution or dealership because the owner fails to make the payments.

Stop and consider for a minute. The car was taken back for good reason: the owner didn't care enough about the machine to make the payments. Do you think he cared enough to perform any service and maintenance over the past three months? Do you think a quart of oil was ever added, much less the pan drained and a fresh supply added along with a filter? Chances are the only thing done was the weekly fill with gas.

Do you think the owner avoided bumps or potholes in the road? Do you think, knowing the car would be repossessed (and most going into the purchase do know that the car will eventually be grabbed away), that he didn't perform some act of sabotage on the car before having to give it up?

I recall the story of a man who refused to make the

payments on a car because mechanically it didn't work right from the day he bought it. He couldn't get the dealer to repair the car and he refused to pay for the repairs out of his own pocket. Told he was stuck with the car, he went out and bought a case of lemons and deposited them in the trunk. The lemons were then sprayed with water and let to sit in the trunk in the hot sun for two months. After awhile, the smell emanating from the trunk was no longer lemon.

The car was repossessed — and had to have the entire trunk lining replaced at considerable expense to the dealer.

That episode points up another reason why cars are repossessed and why it's wise to stay far away from them: the car has a major mechanical problem that can't be fixed and either no one can or will help the owner. The seller got his $15,000 and now wants nothing to do with him.

Or, perhaps, the cost of repair is greater than the value of the car. In that case, the owner has two choices: spend the $1500 to have the repair done and still make the 23 remaining monthly payments at $200 a month or not spend the $1500 on the repair and not make any more of the 23 payments of $200 a month either.

Often there isn't a choice and the car is repossessed. The dealer or financial institution takes it back and says you, the next buyer, can have it if you make the remaining 23 monthly payments of $200. He doesn't mention the $1500 repair that's needed. Tell him, "No thanks."

Recalls are different. A recall can be a good buy or a bad buy depending on what was wrong with it and whether it has been repaired. And with GM and Ford cars, there's a way of finding out. With GM, you call CRIS. With Ford, you head for an OASIS.

Both are automated information systems containing data on each company's recalls. CRIS stands for Computer Recall Identification System and contains data on GM trucks going back to 1973 and on GM cars going back to 1974 or 1978, depending on the division. OASIS stands for Online Automotive Service Information System and holds recall data on all Ford cars and trucks going back to the 1978 model year.

The way it works is this: a dealer is about to take a 1978 GM car in trade. He simply takes the Vehicle Identification Number (VIN) off the dash plate and feeds it into his CRIS computer at the dealership. By hooking up the phone, his computer is able to talk to the CRIS computer in Detroit, which gives him a readout of the recall history of the car— what recalls were issued for that model, what fix needed to be made and whether those repairs were made to this particular car. If not, he can then perform any outstanding recall work before putting the car out on the lot. He'll be reimbursed for the work by GM but don't be surprised if the trade-in value of the car suddenly comes down a bit.

CRIS and OASIS perform a valuable service for dealers. But consumers can take advantage of the systems, too.

Say you're looking at a GM or Ford car on a dealer's lot. Just to be certain, ask the dealer for the CRIS or OASIS data on the car's recall record. He doesn't have it at hand but he assures you everything is OK. Ask him to run the VIN through the system again so you can see for yourself.

If there is an outstanding recall, federal law requires the manufacturer to perform the repair free. So neither you nor the dealer have anything to lose by calling on the computer to perform its magic.

Or, like a lot of people, you plan to buy a used car from a friend, relative or stranger. If it's a GM or Ford car, call or visit a local dealer and ask that the VIN be run through for you. Tell the dealer that if there is an outstanding recall, the car will be brought to him for the work, which means he'll make money for his troubles.

By checking the recall record, you gain a slight bargaining edge with the seller. You are going to have to take the time to have the repair made and you can tell the seller that you expect him to come down somewhat on the asking price.

Of course, the recall data may tell you that you don't want the car at any price. Say the car has been recalled 11 times (the GM X-body compacts, for example). One or two recalls can be tolerated as a simple mistake. Eleven would certainly hint that the engineers and designers had their eyes closed.

Or, the car has been recalled once for brakes because a pad was faulty, once because a transmission part kept breaking, and once because the engine's computer system wasn't working properly resulting in too rich a fuel/air mix, excessive emissions and poor fuel economy.

The CRIS or OASIS system reports each of the three troubles and the fact each of those three items was repaired. No problem. You can breathe a sigh of relief. But what if CRIS or OASIS reports back that the car wasn't brought back for any of the three problems for an inspection? Sure the work can now be done free of charge but you have two pieces of evidence that you don't want the car at any cost.

Clue number one is that by disregarding the recall notices, the owner now trying to sell the car was operating the vehicle under unfavorable if not dangerous conditions. Clue number two is that if the owner had so little respect for that car as to avoid *free* recall repairs, what else has he or she neglected on the car? The owner disregarded brakes, transmission and engine recall notices. Do you think he changed oil or has been running with the proper coolant to keep from damaging the engine? Chances are he or she hasn't.

There are two other ways to use CRIS or OASIS to your advantage. Does your state have an emissions testing program requiring you to bring the car in for annual testing? Then make sure the used car you are about to buy has been run through CRIS or OASIS to check on any emissions recalls. And keep in mind that even if there hasn't been a recall, if you run your car through the emissions test required by your state and the car fails because of a faulty part or component in the emissions system, the automaker is obligated under federal law to make a free repair or replacement for the car's first five years or 50,000 miles of life.

Finally, CRIS or OASIS should be called upon if you've moved and aren't sure if any recall notices have kept up with your change of address. If you find there has been a recall, take the car in for an inspection and have the work done. Keep records to show the next prospective buyer that all work has been done.

There's one catch in what I've just told you about CRIS and OASIS. For the program to work to your advantage it means the dealer must be cooperative. He must say, "Sure, bring your car in or give me the VIN and I'll be glad to find out the car's recall history."

What I've found over the years is that there is a group of dealers who wouldn't turn on CRIS or OASIS for their mother unless there was a dollar to be made in the deal. Some dealers will lie and tell the customer that they have never heard of the system or they don't have one of the terminals in their shop. What I've done is call the local public relations or zone customer service office and ask for the name of a dealer closest to the motorist who has a CRIS system. It doesn't hurt, of course, to call the dealer ahead of time and make an appointment to have someone activate the CRIS or OASIS system for you. If 20 people showed up at a dealer's door on a Saturday morning wanting a recall history on their car, it would be understandable for the dealer to balk on a busy sales day. So call and let the dealer know you're coming.

If you don't own a GM or Ford car, or don't want to take advantage of CRIS or OASIS, you can call the government's safety recall hotline at 1-800-424-9393 to find out if your car has ever been recalled. What you cannot find out is whether your particular car has had the repairs performed. Unlike OASIS and CRIS, the government does not have that information. You'll need the Vehicle Identification Number. You can also use that number to register a complaint on a safety related problem with your car. The government uses those complaints to initiate recall investigations.

CARS TO LOOK OUT FOR
PART 2

Floods, fleets and grays

You pick up your morning newspaper and there on page one is a story of a devastating flood on the East Coast. Damage runs into the millions. Houses have been destroyed. Some people are missing. How sad. You sympathize and then turn the page.

It happens every year. Maybe not always on the East Coast, but somewhere in the U.S. there's a major flood. What has that got to do with cars and buying a used vehicle in Chicago, Milwaukee or Peoria? Often flood-damaged vehicles are shipped out of state and sold as used vehicles — in Chicago, Milwaukee and Peoria.

Many dealers and factories will donate flood damaged cars to high schools or community colleges for use in a mechanics course. The requirement is that the car be used for instruction in service and repair but not be titled to be driven.

But rest assured that while some people are rushing in to help flood victims there are other people rushing out equally fast with cars to peddle elsewhere to try to salvage a sale. The best place to dispose of them is where the disaster has been just an item on the ten o'clock news but not an event experienced in person and where there is an active market in used cars. That makes Chicago and Milwaukee prime targets for dumping.

Remember back when Mount St. Helen's erupted in Washington State? Cars were buried under ashes. Those still driveable soon weren't because air cleaners were clogged with ash. There were reports of ash getting into fuel, oil and coolant systems and requiring costly repairs—if they could be cleaned out and repaired. Within days of the volcanic eruptions cars from that area were spotted on used car lots in the Chicago area.

So what to do?

Beware of used cars for a few months after any disaster. Be especially alert with a very clean late model car with an especially low price tag. Don't be paranoid and assume that every clean late model sold at a low price within three months after a flood is a damaged car. But do take some extra precautions. A dealer I know said when he goes to car auctions and suspects a car has been in a flood, he first conducts his own inspection and asks for the title.

Sometimes the title won't say "submerged" or "flood car," he says.

"But when I know that New Orleans had a foot of rain and the car is titled in Louisiana, I know not to risk it. If the car is titled in any state which has been in the news with a flood problem, I won't buy it. As nice as the car may look at first glance, it just doesn't get any better. You might get most of the mildew smell out of the car, but you'll never stop the rust."

To detect a car that may have been submerged in water, you should:

1. Check for sand, silt or salt deposits under the carpeting in the passenger compartment and in the trunk. Often a stagnant or mildew odor will be a giveaway.

2. Remove the back seat cushion and drop it bottom side down. If the car has been submerged, sand that has dried in the cushion should fall out.

3. Look into the recesses of the car's intake and exhaust manifold and other crevices for clues about sand or silt deposits.

4. Remove the chrome rim from around a headlight and look for sand or silt deposits.

5. Check the electrical system for signs of corrosion. Any copper connection will have turned green after exposure to salt water.

6. Ask for the car's title to see if it has been stamped "submerged" or "flood car."

If the dealer objects to any of your tests, leave. You're probably looking at a future rust buggy.

Another type of "shipped in" car to avoid is what is known as a "gray market car." Gray market cars are the exotic and limited edition European imports advertised for sale through several outfits at prices tens of thousands of dollars lower than what the car sells for here in the states.

The reason for the price differential often is that the cars are hardly the same as those sold here. The U.S. government requires that any import made overseas and sold here conform to all U.S. emissions and safety standards. That means that the auto manufacturers have to comply by adding all the hardware at their plants.

The reason for the savings on gray market cars, their sellers say, is that they are European versions that have had the safety and emissions hardware added later for less money.

What the gray market buyers don't find out until too much later is that the *real* reason for the savings is the slipshod and even dangerous manner in which the equipment was added. Mercedes-Benz, for example, once found that a catalytic converter had been welded beneath one of its cars about an inch from the fuel tank. With converter temperatures reaching 1600 degrees, that car was a mobile bomb.

Other gray market buyers have found that the reason they paid less for a Mercedes was that the car was a *three year old* version of the car they wanted . . . and thought they were getting.

A gray market car cannot be driven on U.S. roads until all the safety and emissions hardware has been added and the car is certified as meeting federal laws. If it isn't done properly, guess what? No car. Most people have found that with the expense of buying the car, shipping it here, converting it to U.S. specs, certifying it and then getting it home, you are better off buying it at the local import dealer.

Another thing—most gray market buyers find that the factory does not offer, much less stand behind, any warranty claims. It breaks, you fix.

So how do you know it's gray market—and you can't tell just from looking at it? The serial numbers on gray market cars differ from the factory numbers. Call your local import zone office to determine whether they differ on the car you're looking at. Check the title. Ask whomever serviced it to check on the records. The mechanic will know. If there's any doubt—Don't.

Another category of cars to look out for are those that come out of fleets. We've already talked about cars from rental fleets and why you shouldn't buy them but there are other fleet cars to avoid as well—such as taxis and police cars.

The cars to be alert for here are Impalas which are frequently used by hack drivers and Dodge Diplomats which are often used in police fleets. And if you don't know why, what can I say?

CARS TO LOOK OUT FOR
PART 3

Chrysler boasts that it sells more turbo-powered cars than anyone else in the industry. And they do. Know why? Because Chrysler has the least number of engine offerings in the industry and has only four cylinder engines available for its front drive models. It *had* to offer a turbo to provide power for its cars that GM and Ford could provide with their wider variety of engines.

The turbo is a crutch—and that's my main objection to it. The turbo became popular during Detroit's downsizing days when the manufacturers were spending so much money developing smaller-sized cars that they didn't have enough left over to bring out new engines. So they simply threw on turbos . . . as a temporary solution. What the presence of a turbo means, in other words, is that the engine isn't powerful enough on its own.

It's also a dirty process. A turbocharger redirects used exhaust to power the turbo and requires more frequent changing of oil and oil filters and therefore more maintenance. And it's hot and without an intercooler you're risking eventual havoc with your turbo and engine.

Turbo-charged cars lead my list—along with the GM diesels—of Lemon Tarts . . . cars that I wouldn't buy, and would advise you not to buy, at any price. With those two at the top, here's the rest of the list:

1. *Any turbo-charged car* — Already discussed in depth. Enough said.

2. *Any GM car with the 350 cubic inch V-8 diesel engine* — A gas engine converted to diesel and it is difficult to pinpoint any part that didn't have a problem. Resale is nil; even dealers don't want them in trade. Since they are so inexpensive to purchase, why not get one and put in a gas engine? To do the job right would take about $5000. You can get a better car for that price.

3. *Any used Vega* — The car that, when new, was a beater. The fenders rust and the engine warps.

4. *Any 1974 model car* — This is the year before the catalytic converter first appeared on cars. Before converters, the automakers had to retard spark plug firing and detune engines so that the engine would perform the task of cleaning up emissions in order to meet federal clean air regulations. As a result, the engines started and stalled — not always in that order — stumbled, coughed and generally got lousy mileage.

5. *The Cadillac Eldorado, DeVille, Seville and Fleetwood* — The 1981 models that tried to bring together V-8, V-6 and four cylinder operation in one engine — unsuccessfully. In the '81 model year, GM was reaching for straws. Federal mileage laws required GM to get greater fuel economy from its cars but the Cadillac division wasn't making a corporate contribution. So to help meet government standards and to try to lure buyers to a car associated with gas guzzling when gas prices were high and supplies tight, GM came up with the idea of a variable displacement engine that could run on four, six or eight cylinders. Need power to pass? Just hit

the pedal and eight cylinders would let 300 horses loose under the hood. But cruise at highway speed on the interstate and only four cylinders were needed to provide the impetus. Great concept — dreadful execution. The controls were never fully in sync with the engine and the driver never knew if he was going to get the V-8, V-6 or four cylinder mode. The engine lasted one year in Cadillac cars, a few more years in its limos.

6. *The 1982 Chevrolet Cavalier and Pontiac J2000* — Don't ever race a Schwinn bicycle in one of these cars. This was the first year on the market for the J-body subcompacts and the cars looked nice enough but the problem was the engine — a 1.8 liter carbureted plant that lacked the power to accelerate unless going downhill with a stiff tailwind. Even then, merging required a very heavy foot and a handy rosary. Designed three years earlier with the belief that gas would reach $2 to $3 a gallon, the carbureted engine debuted in cars in the spring of 1981 as '82s when gas prices were $1 a gallon, supplies were ample and customers were looking for performance rather than mileage.

7. *Any 1985 Buick Electra, Olds 98, Cadillac DeVille* — Another first year car, this time the first of the downsized, front-wheel drive versions of the old full size, rear drive models. The cars were delayed from being introduced as '84s as GM scrambled to rectify transmission problems. GM, in fact, without making a very big or well promoted deal out of it, was replacing transmissions in early run models, an obvious indication that not all the problems were worked out ahead of time despite the delayed introduction. Early cars had another minor problem — glare off the dashboard into the eyes of

tall motorists. A fix was made in the form of a plastic lip on the dash to eliminate glare. The transmission problems alone would scare me away. It's one thing to be a silly goose, another to be a guinea pig.

8. *Any 1975-1979 Honda Civic or Accord* — Rust problems galore.

9. *Any 1980-1981 Honda Prelude* — Ditto the rust.

10. *Any 1976-1977 Dodge Aspen or Plymouth Volare* — Rusty fenders, poor performance, lousy mileage. You wonder why Chrysler almost went bankrupt in 1979-1980? The cars it sold prior to that certainly helped. In fact, add most Chrysler, Dodge and Plymouth cars built before 1980 to the list of questionable used cars today.

11. *Any American Motors Corporation Alliance or Encore* — Distress merchandise, even the new ones. When Renault took over AMC some quipsters called the new partnership FrancoAmerican. The company that markets spaghetti under that name must have felt bad.

12. *Any Renault LeCar, 18i sedan or wagon, or Fuego* — All French imports from Renault sold by AMC. France makes good wine, some suspect cars.

13. *Dodge Omni or Plymouth Horizon* — The engine is what to look out for here. Chrysler used the 1.7 liter four cylinder from Volkswagen from 1978 to 1982, the 1.6 liter four from Peugeot in 1983 through 1986. Avoid cars with those engines. Best bet with these cars is the 2.2 liter four cylinder built by Chrysler itself, which means you can find all the repair and replacement parts you need without relying on VW or Peugeot. Most important, the 2.2 is a good engine and best suited for these cars.

14. *Any Chrysler car powered by the 2.6 liter four cylinder engine supplied by Mitsubishi* — A temporary engine also offered in the 1984–1986 Dodge Caravan and Plymouth Voyager mini vans. Another stopgap engine used by Chrysler until it was able to design and build its own 2.5 liter engine first offered in the '86 model year in cars but not vans. Getting parts is one problem. And if the engine was that great, do you think Chrysler would bother spending $50 million to come up with one of its own? When an engine is temporary, finding parts years later is a chore — and costly.

15. *Any 1980 GM X-body, front-wheel drive compact (Chevy Citation, Buick Skylark, Olds Omega, Pontiac Phoenix)* — Poor engineering, poor assembly. Main charge is that the brakes lock in a panic stop. Maybe they do, maybe they don't, but finding out could be a painful experience if the answer is maybe they do. But even if the brakes were perfect, a close exam will show shoddy workmanship and assembly: doors that don't line up, body panels that don't match front to rear. I owned an '80. Within the first month the left rear spring broke — sitting in the garage. Note that later models got progressively better as more attention was paid to detail. Prices usually are at the giveaway level if you're willing to accept poor resale.

16. *Any Pinto* — Allegedly the car could become an oven in a rear-end collision because the gas tank and axle were placed too close to one another.

17. *Any Fiat* — For every weekend you spend driving a 124, 128, X 1/9 or Strada, you'll spend three weekends trying to get it running again. A friend once said: "They are great in motion but you spend most

of your time with them in commotion." A car similar to any Fiat would be any Triumph.

18. *Any 1980 or older Jaguar*—John Egan, who took over Jaguar in 1980, once told me: "We sold only 3000 cars in 1980 and none of them were any good." Egan obviously would never make it writing car ads in Detroit.

19. *Any Audi*—Yuppymobile. Other than Volvo (except 760 and 740 GLE), the most overrated cars on the road. The 5000S is under investigation for sudden acceleration when shifting from park to drive or reverse. Repairs, service and maintenance are all rather costly.

20. *1986 Yugo*—The only car in recent memory that when new made buying used a better choice. An offshoot of the Fiat 128, the car was introduced in 1986 and sold well at first based on a $3990 price tag, lowest in the industry. Other than not being a very good car, the drawbacks include sales of about 40,000 units, hardly a number that will make finding parts easy down the road. Government crash test results claim that in a 35 mph impact, the cost of repair would be almost the cost of a new car.

21. *The 1972-1979 Ford and Mercury rear drive cars*—There are claims that their transmissions jump out of park into reverse on their own. Another situation where the allegations may prove false but who wants to find out the hard way? The federal government ordered Ford to send consumers gummed labels to display on the dash of these cars warning motorists to be careful when putting the car in park to insure that they indeed hit park and not reverse. But it's difficult to fathom why a nation that refuses to wear seat belts would bother with gummed labels.

22. *The 1974–1978 Ford Mustang* — Small, heavy, bulbous cars that didn't do justice to the sleek Mustang image of the past and got incredibly poor mileage.

23. *AMC Pacer* — The saucer or bubble car which was supposed to house the Wankel rotary engine General Motors was going to sell AMC. AMC built the Pacer from 1975 to 1980 but GM never built a rotary. Too bad.

24. *The 1981–1983 Chrysler Imperial* — This was simply the Chrysler Cordoba with a bustle-back, protruding rear end for several thousand dollars more than a Cordoba. A sales disaster. Chrysler will resurrect the Imperial in 1989.

25. *The 1982–1985 Ford EXP or Mercury LN-7* — These cars were Ford's experiments with two-seater sporty cars. The first problem is styling — they just don't look very good, especially when matched against the Pontiac Fiero or Toyota MR-2. The second is the engines — they are underpowered, being Ford's 1.6 liter carbureted mileage model. Ford came back in 1986 with the EXP with a more powerful 1.9 liter engine, but the styling is still lackluster. The LN-7 was never brought back. If you like Ford and want a car that combines sportiness with high mileage the Escort GT is the choice. Very nice.

26. *Isuzu I-Mark* — I once asked a variety of car dealers to name the vehicle offered to them in trade that most made them wince, the one which was most likely to sit on their lots gathering dust. The response in all but one case was the I-Mark.

27. *Any GM car powered by the 3 liter carbureted* V-6 *engine* — Next to the diesel, GM's worst engine ever. It should not be confused with the 3 liter *fuel injected* V-6 GM put into its 1985 model N-body

compact Grand Am, Olds Calais, Buick Somerset and Skylark. The carbureted engine went into a variety of mid and full size '83–'85 cars but mainly into Buicks because it was a Buick engine. It had a reputation for start and stall, surge and delivering poor mileage.

28. *Any 1976 to date Chevrolet Chevette* — Cheap — with good reason. It is small, noisy, cramped and sluggish. It is a favorite of rental companies which is another good reason to avoid it.

29. *Pontiac Fiero* — The two-seater is very stylish but the ones I'd stay away from are any with the base 2.5 liter engine and manual transmission, especially the 1984 models which were the first ones introduced. With the 2.5 and manual you have a car built for show, but no go. The optional 2.8 liter V-6 has the power needed and automatic is always preferable in a GM car. The manual in the Fiero is one of the sloppiest shifting in the industry. GM just doesn't make a smooth-shifting manual.

There they are . . . the leading Lemon Tarts of our age. It's obvious when you look over the list that many of the cars made it on the basis of the engine and/or transmission. The 1983 Olds 98, for example, is a really nice car. It has all the room, comfort, power and ease of maintenance you could ask for. But put a diesel engine in it and it becomes a mediocre machine not to be considered unless you have a masochistic streak in you and love spending time in the repair shop.

As a general rule, I would take a fuel injected engine over a carbureted engine when shopping for a used car, excluding of course the first year that fuel injection was offered on any make. With fuel injection and computer controls, you get faster, surer starts, precise fuel-air mix metering, high economy and lower emissions. A fuel injected engine, thanks to the computer, will compensate for such variables as humidity or altitude when you're going up a mountain. A carbureted engine will not.

CARS TO LOOK FOR
PART 2

Cream puffs

So are there no good cars on the road? Yes, there are — and this is where I name some. This is my list of cream puffs — the cars that I would consider buying . . . and why.

A word of caution. We're talking here about models, not individual cars. A car might start out in life with good genes but be subject to abuse. It is necessary in every case to pass judgment on the *individual* car. But you get a head start by knowing which cars are potentially good and which are inherently bad, such as the ones on the list of Lemon Tarts.

Keeping that in mind, let's get on with the list.

1. *Toyota Mr-2* — Super car. 5-speed best.

2. *Toyota Celica* — Wait for the newly styled 1986 models to reach the used car market. After 16 years, Celica put its act together.

3. *Toyota Cressida/Nissan Maxima* — When you're very good new, you're very good used. Remember what I said about track record.

4. *Dodge Shelby Charger* — Among the sportiest looking cars made since 1983. I especially like the silver and blue two-tones. Except for the '84 model, when automatic was offered as an option, you'll find manuals only.

5. *Ford LTD/Mercury Marquis* — Midsize, rear drive models built between 1983–1985 as successors to the Granada/Monarch were dropped in 1986 to make room for Taurus and Sable. Conservative styling but reliable transportation with good size and room.

6. *Ford Mustang* — Buy 1979 or later models to avoid the bulbous 1974–1978 monstrosities (remember Lemon Tarts?). Really sharp looks and adequate performance but you'll have to live with modest fuel economy at best.

7. *Chevy Cavalier* — With the addition of the 2 liter, four cylinder engine in 1983, this is now a decent performer. The optional 2.8 liter V-6 offers a bit more zip.

8. *Any model Chevy Camaro or Pontiac Firebird* — These are among the most cramped, hard-riding, lackluster performing cars in the GM stable. But styling stands out. Teens love them so you'll have no problem getting rid of one once your kidneys and bladder have had enough of the "youth experience." Demand is higher for used Camaros and Firebirds than for new ones. One reason is that the youths who love the cars can't afford them new so they have to wait five years until they can manage the used car price tag. Don't waste your time or energy on a rough-shifting manual. Go automatic.

9. *Any year Chevy Corvette* — These are among the most cramped, hard-riding, lackluster performing . . . oops, I already said that, didn't I? The 'Vette isn't a car for those who want an air-cushion soft ride. The 'Vette is a car to be seen in. Passers-by (except those who own one) don't realize what abuse the 'Vette driver is putting up with. The beauty is that the car is an investment and often will return

you as much money when you eventually sell it as you paid for it. Special caution: cost of replacement exhausts, tires and tune-up are extremely high if you should find one in need of all three. If you've never dealt with 'Vettes before, make friends with a club member (easy to do) first and go slowly before making the leap. Stay away from the awkward-shifting manual.

10. *1983 and newer Chevy Celebrity* — The 2.5 liter four cylinder will give you good mileage at the expense of noisy acceleration. The 2.8 will give you quieter acceleration at the expense of good mileage.

11. *Chevy Impala/Caprice* — Watch out for old taxis but other than that you'll get great room and comfort and seating for six adults inside and their luggage or your groceries in the trunk. Been around so long the bugs are gone. You can find a mechanic almost anywhere to work on one. And you can find replacement parts almost anywhere.

12. *1985 and newer Pontiac Grand Am* — Acts more like a sports coupe than an economy car. One of the hottest cars at Pontiac new and should be in demand used. Best bet is with the 3 liter V-6 fuel injected engine. Two-door models look best and four-door models don't offer that much more ease of entry or exit in back. Very good performance and at the same time very good mileage.

13. *Buick Century/Olds Ciera* — Same as Celebrity. Best bet is the 3.8 liter V-6. If the car has the 3 liter carbureted engine, walk away.

14. *Any pre-1986 downsized, restyled model Buick Riviera/Olds Toronado/Cadillac Eldorado* — The old ones have character, yacht-like size, trunks as big as basements and rear seats that hold three adults and

are getting hard to find. The new ones all look like clones. Offers same benefits as Impala/Caprice in terms of service and problems worked out over the years.

15. *Any pre-1986 downsized, restyled, front-wheel drive model Buick LeSabre/Olds 88*–The newer LeSabre/ 88s look just like the newer Electra and 98, which basically is bland. Offers same benefits as Impala/ Caprice on service and problems. Ditto on warning about 3 liter V-6 carbureted engine.

16. *Olds Cutlass Supreme/Buick Regal*—Remnants of big, rear drive cars. Cutlass especially distinctive with Regal close behind. Comfort and room and proven track record are the reasons to buy. But, yep, watch out for the 3 liter carbureted engine.

17. *Any 1980–1985 model Cadillac Seville with the bustle-back deck lid*—I hated the looks of that rear end—and then I saw the sterile new 1986 model. The ride is really mushy, luxury type boulevard style.

You will note that, with the exception of the top three, every car on the list is a domestic. That's because, as I've said before, unless you live in California it's much easier to get a three to five year or older U.S. built car serviced and repaired than it is an import.

You'll notice something else about the list: there's not a rare bird on it, not a single reference to something exotic like a Ferrari or cutesy pie like a Triumph TR-7. I can't emphasize enough that if the car was a low volume number new, it becomes even harder to find engine or body parts when used.

Every one of my recommendations for cream puffs is a car that you can get serviced or repaired without having to make an appointment with a high-priced specialist, though when it comes to the Corvette I wouldn't pull into the corner gas station and have old "Tom Thumbs" do the tune-up.

Value for the dollar doesn't simply refer to the price you *pay* for the car. Value refers also to the cost of *maintaining* and *repairing* the car years after the new car smell has disappeared.

WHERE TO LOOK

Now you know in general terms what kind of a car to look for. Where do you find it? There are a variety of places to find used cars. There's the new car dealer who sells trade-ins as used cars. There are the corner lots that hang hundreds of plastic flags from the telephone poles and deal only in used cars. And there's the friend, relative, neighbor or some other private party who has a car sitting in the driveway with a sign on it or who has put an ad in the local paper.

According to the most recent survey done by the Hertz Corporation, the rental car folks who annually take a look at used car buying trends, 37 percent of all used cars are bought through new car dealerships, 12 percent at those corner lots and 51 percent through private party sales, which include relatives.

It used to be that the new car dealer was absolutely the place to find the best used cars. He had a reputation to maintain and so he would keep only the best of the trade-ins and dispose of the rest, either by farming them off to the weekly dealer auctions where the corner lot owners would buy them, or wholesaling them for junk or sending them to the South.

That isn't true today and the reason is found in another Hertz survey figure—the number of years that owners are keeping cars. Typically, Hertz reports, the length of ownership today is 4.6 years, up from 4.5 years in 1984 and up dramatically from 2.8 years in 1979.

Owners are holding on to cars longer. As a result, when they are brought in for trade on a new car they have greater mileage and far greater wear, physically and mechanically. Rather than disposing of these cars at the auction or the corner lots, dealers are holding on to them in order to have used cars to sell. That means that the dealer's lot is no longer necessarily the best place to find the right used car.

Most people are shopping among friends, relatives and private parties. People are cautious about buying used cars and if they know of an aunt, an uncle or a friend who has been taking good care of the car, they don't hesitate to take it off their hands. I wouldn't either. Or there's a relative or a neighbor who spends every weekend working on his car, tuning it up, changing the oil and the like. You know the car is in great shape so put out the word that when he is ready to buy again, you'll be happy to take the car off his hands. It'll be a wise move.

As for the corner lots, they are rarely a good place for buying a used car because of the way they get their cars — usually from the weekly dealer auctions. The cars they get often are the ones with fairly respectable looking bodies but with engines and transmissions on their last miles. The dealer dumped the car at auction because he didn't want to spend the time or money reconditioning it to sell on his own lot so the plastic flag man picked it up cheap.

The plastic flag lot is a good place to pick up the repo, the rental with 35,000 miles usage in six months or the car hanging on for a few more miles of life by a thread.

Here you will find the accident victims, the cars whose bodies weigh heavily with plastic filler, the ones a dealer wouldn't take for fear they had something catching. The prices here are often hundreds of dollars less than you will find at any dealer lot — and for good reason: the cars are worth at least hundreds of dollars less than those at the dealership. You could be buying a car that will be needing a major repair soon and many minor ones after that.

Dealerships are equipped to perform warranty work on used cars, if a warranty is offered. But the carnival-style flag lots simply sell cars. They seldom service and they usually don't warrant what they sell. If there is a shop on the grounds, it usually is a paint facility to cover rust or a room where a buffer is working around-the-clock to wax life into an over-the-hill machine.

The carnival lots are the flea markets of the auto industry. You must be desperate to try them.

Don't expect a warranty in a private sale, either. If you buy from a friend who lavishes care on his car and it breaks down — well, that could be the end of a beautiful friendship. Keep that in mind.

And don't expect wrongs to be righted if you buy from a previously unknown private party. The thing to look for here is not a warranty but a clear title to the car to prove that it belongs to the person and wasn't stolen — and a Vehicle Identification Number on the dashboard. If the number is missing or looks altered, chances are you are looking at a stolen car. Run.

Whether private party, flea flicker or dealership salesperson, what the person selling the car has in mind is one thing — to get you to buy the car for the maximum amount of money they can get from you. And the more experienced and professional the seller is, the more adept he or she is likely to be at doing that.

Remember that cup of coffee you got offered when you walked through the door of the car dealer's showroom back in the Introduction? Nice friendly gesture that, wasn't it? Sure was. That salesman is so anxious for the pleasure of your company that he has just guaranteed that he is going to have it for the next half hour.

Try taking a sip of that coffee. Make it a little sip. Make it a big sip and you are going to be speaking sign language for the rest of your life. That coffee is so hot that it will burn your tongue out. So you put it down for awhile. On that desk over there. And while it cools you and the salesperson walk around the lot, looking at the used cars. Twenty minutes later, you come back to claim your coffee and you find that you just happened to put it down on the same desk where the contract for the car he wants to sell you happens to be.

That's just one of the little ploys that car dealers and their salespersons have for separating you from your money. The process is littered with them, like land mines, right up to the point when you sign on the dotted line — and even beyond that if you let yourself get talked into a service contract. We will point them out to you as we move you along through the process.

Some are subtle, like the hot cup of coffee routine. Or like locating the rest rooms or pay telephone deep in the heart of the dealership. Why? Well, for one thing it forces you to walk past all those nice shiny new cars and vans, both on the way in and on the way back out.

It also gives the salesperson more time to work on you. Guess who is going to be waiting for you when you walk back out the door of the rest room or phone booth, ready to pick up the discussion on the merits of the car he or she wants to sell you?

In any car buying transaction, time is on the side of the salesperson. The more time he or she gets, the greater the likelihood of making a sale. The more time you wait, the more likely you are to give in.

Sometimes, giving in isn't left up to you. There's a story told of a former Chicago-area car dealer who would wait until the potential customer got into the showroom and then have one of his car hikers pull a used car behind the visitor's car and then walk away. When the customer was ready to leave, the staff would have to run around looking for the person who had blocked his retreat. Meanwhile, the salesman had a few more minutes to try to make a sale.

As I said before, there is one thing that the salesman wants and that's your money — as much of it as he can get. That's what the car-selling business is all about and getting you to part with it is what he is an expert at . . . or he wouldn't be in the business.

So is there nothing that you, as a buyer, can do to protect yourself?

Yes, there is. In fact, there are several things.

The first is to take the advice that we've offered in this chapter and be selective about who you choose to buy from. Stay away from the flea flickers and any other dealers — or private party — that you suspect might not be quite on the square. Ask around and find out who has the good reputation in town. Usually this will be a dealer who has been in business for awhile and values his customers and his reputation. Car selling, after all, is a repeat business and good dealers want their customers coming back to them.

The second thing is that you now have some minimal protection built into the law. It is not what it ought to be and it is not what it started out to be but it is, at least, better than it used to be. We'll discuss that in the next chapter.

Third, you can give a car a thorough going-over, not only personally but with a mechanic. If the mechanic is any good, you should pretty well know what you're getting. If you come up against a dealer who won't let you bring a mechanic in to look the car over, walk away from him. You're dealing with the wrong one.

Finally, let me say one other thing. There is basically nothing wrong or dishonest about a dealer making enough money to pay his bills and have enough left over to live the good life or for the salesman to earn enough commission to feed and clothe his family — provided you get fair return in the form of a vehicle that will take you to work, to the store, the church, the school, the mall or on vacation or whatever. That, after all, is what you went out shopping for in the first place.

REGULATIONS

Some new rules of the game

In 1985, after years of so-called deliberation and modification, the Federal Trade Commission finally adopted a Used Car Rule calling for new guidelines in the sale of used cars.

I say "so-called" deliberation and modification because for the most part the government and nation's car dealers actually spent years watering down the original proposal.

What the staff of the FTC had originally wanted was for used car dealers to inspect each and every car that came in on trade, make whatever repairs were needed and then list on a window sticker what those repairs were. The thinking was that the potential buyer just might like to know what had gone wrong with the car and what was done to fix it before he or she plunked down the money.

The nation's dealers went into a fit over that proposal. After the best lobbying job that money could buy, a bland rule finally was agreed upon that made the dealers happy.

What the dealers objected to was (1) having to inspect the cars, which would take time and, worse, cost money. And if the mechanic did find something wrong he would (2) have to fix it, which for sure would take up valuable mechanic time and extremely valuable dealer profit. But worst of all, the dealer (3) would have to actually tell the customer about a problem by listing it on a window sticker.

It would be, dealers argued, like having to pay to go to confession.

What the FTC finally came up with is a "Buyer's Guide" window sticker for display on all the cars on the dealer's used car lot. The sticker tells the potential buyers just about everything they ever wanted to know about the car — except what, if anything, is wrong with it.

The window sticker requirement is far short of the original used car proposal that would have made it mandatory for dealers to inspect and repair trade-ins before they were put on used car lots. But the car-labeling guides do provide some information for prospective buyers that in the past either was relegated to the fine print or was made in the form of verbal promises that were seldom kept.

The National Automobile Dealers Association, which represents more than 20,000 of the nation's new car dealers, devised the labels, which is like having the judge ask the defendant to come up with his own sentence. But since the used car rule had been debated for so many years, the FTC considered that anything was probably better than nothing.

The labels vary somewhat in appearance in various states so I'll go by the one adopted by the Chicago Automobile Trade Association which represents more than 750 Chicago-area new car dealers and which is representative of what you'll find elsewhere. (See example at the end of this chapter.)

The label contains three key provisions that hadn't been associated with used car sales before and in that respect it serves a useful purpose.

The first provision is a statement that tells the potential used car buyer to "Ask the dealer if you may have this vehicle inspected by your mechanic, either on or off the lot." This is a substitute for the original proposed wording which was: "Arrangements may be made to have this vehicle inspected by the mechanic of your choice." Words to dilute by.

But even the statement that exists is a 100 percent swing

from the way most used car dealers had done business. The reason it got changed was that the used car dealers didn't want the bother — or the liability — of checking out the car for problems so they have placed that burden on the consumer. While that means that the consumer must find a good mechanic to perform the inspection and then pay for his or her time, at least it gives the buyer the option of trying to find troubles before the purchase and not discovering them by accident a month later.

The second provision is a note regarding those oft-made, seldom-kept verbal promises. "Spoken promises are difficult to enforce. Ask the dealer to put all promises in writing." Again, as originally worded, that statement had read: "Spoken promises can lead to misunderstanding. We will gladly put all promises in writing." Not quite the same, but the provision as it exists does give you the opening to ask for all promises to be put in writing. And if the dealer doesn't want to do it, walk away.

The label guide also contains statements and boxes to be checked that inform the shopper that the car either does or does not have a warranty and is being sold "As Is." If it does have a warranty, there's an explanation as to whether it's a full or limited warranty and, if limited, what the dealer will pay. There also is space for the dealer to list what car systems are covered by the warranty and for what period of time.

There also is an explanation on the label that if a service contract is available on the vehicle that this is an extra charge item. And on the back of the label guide is a list identifying "some major defects that may occur in used motor vehicles." (See p. 73.)

This really is a pathetic feature. The fact that a service contract is mentioned on this official government form makes it appear that the poor dealer really doesn't want to be able to charge you $200 or $300 for a service contract but big, bad old government is making him do it.

That provision must have the dealers laughing all the

way to the bank. They didn't want to spend the time or money to inspect for and fix repair items on the car but they sure are willing to sell a service contract they stand to profit from.

"It's a used car and something can go wrong with it and in the long run you'll save money if something major goes wrong," dealers always tell me.

"If you inspected the car and replaced the brakes when they have only three miles worth of lining left on them, the buyer wouldn't have to worry about the major repair," I reply.

Read the fine print on the service contract. You may find more items excluded than actually come with the car. And you may find you have to perform a rigorous maintenance routine which if you do reduces the chances of a problem developing. There may also be a deductible per visit, such as $25 or $50 up front on *any* repair made. Or, and this is typically the clincher, there may be that line that states that the dealer is absolved of making a repair free of charge if there is evidence of owner abuse. To some dealers, owner abuse starts the minute you've bought the contract and leave the lot.

"You brake too hard," customers have been told, "and that's why the linings are shot. That's owner abuse."

"But I bought the car just last week," the customer replies—to no avail.

Whether it's a new car or used car, a washing machine or TV set, when the basis of trying to make you part with your money is to convince you that you're saving money, beware. When the salesman uses the old scare: "But what if something major happens . . ." or "But you just paid $5000 and you *don't want to protect* that investment?" you can rest assured that it's not your peace of mind that he's concerned about but the piece of your pocketbook that will represent his commission and the dealer's profit.

As for that list on the back of the guide telling you of

major defects that could occur, the sharp salesman will point these out as a reason for buying a service contract. Again, the dealership can list all the problems that *could* happen with a car but the law doesn't require him to list any of the troubles that *did* occur with this car and what the dealer did about them. Sensible? Sure, if you're the dealer.

Of all the information on the label, the provision I find to be the most important is the one that informs the customer if the car is being sold "As Is."

This has been a nightmare for car buyers in the past. Too often, a sharp looking used car being offered at an attractive price carried an "As Is" warning in such small print that you couldn't find it, much less read it.

"As Is" means just exactly what it says: what you see is what you get — nothing more.

As the label states, "As Is" means "You pay all costs for any repairs. The dealer assumes no responsibility for any repairs regardless of any oral statements about the vehicle." The originally proposed wording was more specific, stating "As Is" meant: "We make no promises about the condition of this vehicle and accept no responsibility for any repairs after the time of sale. No warranties either expressed or implied are offered on this vehicle. This vehicle is sold 'As Is'. 'As Is' means that you will pay all costs to repair things after the time of sale."

Typically, consumers thought "As Is" referred to the car's appearance, meaning that if it had a few nicks and scratches, you had to live with them. Salesmen did little to discourage customers from thinking just that.

Only after the buyer signed on the dotted line, paid the money, turned on the ignition key and panicked when the engine started smoking and the transmission decided to take a nap, did he or she find out what "As Is" really meant.

With the new labels, the customer at least has a warning that if anything goes wrong he or she will pay for the repairs and the dealer has absolved himself of all responsibility. With the "As Is" explanation, the salesman can no longer say: "And if you find something wrong, just bring it back and we'll fix it," because you'll know it's a con—unless he's willing to put it in writing.

Keep in mind, however, that the simple addition of a label doesn't mean that now all the cars you'll find on the used car lot are cream puffs. The labeling has no effect on the condition of the cars to be sold. It doesn't promise that used cars are going to be better, more reliable, durable or dependable. It simply means, as Ross Kelsey, executive vice president of the Chicago Automobile Trade Association told me: "Too often a person buys a used car, finds something wrong, takes it back and now no one knows him. With the labels, everything is spelled out in advance."

Not everything, as we've seen. It's not a perfect system. But it's a lot better than the label-less windows we had.

BUYERS GUIDE

IMPORTANT: Spoken promises are difficult to enforce. Ask the dealer to put all promises in writing. Keep this form.

VEHICLE MAKE   MODEL   YEAR   VIN NUMBER

DEALER STOCK NUMBER (Optional)

WARRANTIES FOR THIS VEHICLE:

☐ **AS IS-NO WARRANTY**

YOU WILL PAY ALL COSTS FOR ANY REPAIRS. The dealer assumes no responsibility for any repairs regardless of any oral statements about the vehicle.

☐ **WARRANTY**

☐ FULL ☐ LIMITED **WARRANTY. The dealer will pay _____% of the labor and _____% of the parts for the covered systems that fail during the warranty period. Ask the dealer for a copy of the warranty document for a full explanation of warranty coverage, exclusions, and the dealer's repair obligations. Under state law, ''Implied warranties'' may give you even more rights.**

SYSTEMS COVERED: **DURATION:**

_____   _____
_____   _____
_____   _____
_____   _____
_____   _____
_____   _____
_____   _____
_____   _____

☐ **SERVICE CONTRACT. A service contract is available at an extra charge on this vehicle. Ask for details as to coverage, deductible, price, and exclusions. If you buy a service contract within 90 days of the time of sale, state law ''implied warranties'' may give you additional rights.**

PRE PURCHASE INSPECTION: ASK THE DEALER IF YOU MAY HAVE THIS VEHICLE INSPECTED BY YOUR MECHANIC EITHER ON OR OFF THE LOT.

SEE THE BACK OF THIS FORM for important additional information, including a list of some major defects that may occur in used motor vehicles.

THE INSPECTION
PART 1

The once over and once under

You walk into the dealer showroom. Around you are all the nice shiny new cars. The salesman approaches you and you tell him that you want to look at the used cars. He nods and leads you past all the nice shiny new cars and out through a door to the used car lot out back.

Culture Shock! Welcome to the Land of the Used Car.

The cars out here are *not* so nice and shiny and new. Some, in fact, are showing decided signs of wear and tear.

You've just come up against the first fact of used car life: Used cars are used. Someone has already worn all the nice and shiny and new off of them. What you are going to get is a car with some wear and tear, perhaps a nick or two, or some scratches or dirt.

In fact, if you don't find some signs of wear you should start worrying. As with everything else in life, if it looks too good to be true, it usually is. (Remember those flood cars?)

Don't give up on a car simply because you spot some body damage or a fender that's been replaced. One reason you're looking at a used car is to save money and the repaired fender you spot is a negotiating tool to save even more. If this is a young person's first car, chances are there will be some more dented fenders along the way. If it is a commuter car for the train or to run Dad to a construction job, don't let one fender stop you from buying a car that otherwise might be mechanically sound and ready to provide you with three years of low cost motoring.

What you want to be certain of is that the car *is* basically sound structurally and mechanically—that the frame wasn't knocked sideways in a serious accident or the engine or transmission wasn't burned out for lack of fluid.

How do you do that? By conducting an inspection of the car, first on your own and then, preferably, with a trusted mechanic. In conducting your inspection, be alert to negatives but don't let them blind you to the positives.

The first rule is always to give the used car a thorough inspection in the daylight. Darkness, like rain, can camouflage damage. You may want to also test the car at night to check nighttime visibility and how good the lights are and also in the rain to check for leaks. But the primary inspection should be done in the daytime.

The second thing you want to do is to always take along a magnet. This is not for entertaining the kids but to give the body of the car a once-over to detect accident damage. If the body has been filled in with plastic filler, the magnet will let you know. Run it along the rocker panels (lower body strip under the doors), in the wheel wells and along the fenders and doors. If it doesn't stick to the body, you know that there's a wad of plastic filler underneath—and the plastic filler is probably hiding a bunch of rust to boot.

Third, get under the car. Sometimes the body shines brightly, the chrome glistens, the glass sparkles—but the undercarriage is two inches thick with rust because the owner's concern was in making the car look good for sale.

Once it gets started, rust can't be stopped. And if the car is loaded with it underneath, that brilliant red exterior finish will be pockmarked with it within a year.

While you're underneath, check for weld marks that would indicate accident repair. As more cars go to unibody structures and away from frames, accident repair becomes more difficult to detect. But look for a bent frame or signs of metal stress or stretching that indicate the unibody was put up on a straightening device.

With the body-on-frame cars, if you were struck up front, the frame would bend up front. The fender replacement two shades darker than the rest of the body is a give-away to damage. Slide right under the car and look at the frame.

But with unibody cars, the chances are that if the car is kissed in the left front fender, the damage could show up in the right rear quarter panel. With unibodies, you have to inspect more cautiously for damage.

If you suspect that the car might have been in an accident and the damage might have been severe, take the car out to a dry straight stretch of highway. Pour some water on the roadway. Then with the car aimed in a straight line, drive through the water and beyond for several feet. Get out and look at your tracks. Did the rear tires follow in the same path as the front ones or was one going east and the other south or northeast, indicating an alignment problem or perhaps a badly bent frame?

Check the inner door panels for signs that that beautiful red coupe you're looking at wasn't once a beautiful yellow coupe. Repainting the major exposed body panels is easy. Getting at the hard-to-reach inner panels isn't and telltale signs may have been left behind.

Why was the car repainted? Usually, the reason is because it was severely damaged and the owner used the occasion to change colors. Or the body was so rusty the owner thought a cheap, bright cover-up might help him to peddle the car to someone else.

Check the door handles. Are they tight? Do the doors close without leaving gaps? These might hint at a previous accident, or they could be sprung from abuse.

Move to the back of the car and check out the trunk lid and bumper. Is the trunk lid properly aligned with the fenders and the spacing between the two even and without wide gaps? Poor alignment or uneven spacing could mean the car has taken a bad hit.

Look at the bumper. Does it show signs of having once sported a hitch for a trailer? If so, that may mean that the engine, transmission and brakes were subjected to heavy loads for long periods of time. The seller may be getting rid of the car because it can't handle the burden anymore. Even if the car has only 20,000 miles on it, those could be 20,000 tough miles.

Open and close the trunk lid. How does it work? Easily—or does it come crashing down like a guillotine? (Trunk lids, by the way, are usually called deck lids these days for the same reason that the glove compartment has become the glove box and Cadillac dealers call their used cars "pre-driven.")

Now check the inside of the trunk. Will it handle the week's groceries or all the luggage you take on a vacation trip? Where is the spare tire, jack and lug wrench? And the instructions on how to use them? Is the spare tire the same size and construction as the four tires on the car? If you have four steel-belted radials on the car and the spare is a bias ply, handling will be adversely affected if you have to use it. And if the four tires on the car are 14 inch versions but the spare is a 15 inch you won't be able to use it.

Take out the lug wrench and make sure it fits the lugs on all four wheels. Often the original lug wrench and jack will have been misplaced and any old set is thrown in the trunk for sale purposes. You buy the car and maybe a year later you have need for a tire change and you suddenly find you are stuck with a jack that won't work on your car.

Now move around to the front and check the hood for alignment with the fenders and panels. Raise it and lower it to test ease of operation. Look under the hood for signs that the car may have been repainted. The inside of the hood is usually a dead giveaway since there are so many crevices for the painter to cover and most painters miss a few.

If the vehicle you're looking at is a pickup truck or a four-wheel-drive utility vehicle, check the front bumper for signs that a plow might have been attached. And in the case of a pickup truck, check the cargo bed for evidence that a salt spreader was aboard. You'll probably see the rust marks swelling up.

Check the insulating material. Are there signs that the engine heat has turned it brittle or is the insulation missing in spots? In both cases, it will have to be replaced. Or is this a five year old car and the insulation is new? Why? Was it because the hood was replaced after an accident?

Perhaps the seller will watch as you pay special attention under the hood and start offering you such information as "It's a 350 four-barrel and has a 3:20 rear axle ratio . . ."

He may rattle off the information to be a help, figuring you'll be impressed with a 350 V-8. He may also offer the information and step on his tongue in the process, such as: "And it's got a hemi engine with racing lifters . . ." which means the car wasn't used to visit the grocery store on weekends.

But the owner may begin his litany because he's observed that you know little about cars. He learned when you asked: "What year is it?" or when you lifted the hood and asked: "Is it an eight or a six?" (One way to tell is to count the spark plugs. There is one for each cylinder.)

Knowing you're a novice, the seller may try to intimidate you or get you to close the hood and be impressed with the wash and wax when the engine compartment is loaded with clues that will tell you you are heading for trouble.

Look at the battery. Is it loaded with corrosion? You may see some ash-like deposits at the terminals. Not good, but not uncommon. But if you see corrosion or stains at the caps you have evidence that you'll soon have to pay for a replacement at $40 and up. And you have to wonder why the owner neglected replacing it. What else did he neglect?

If the battery is loaded with corrosion, chances are the

tray holding the battery has been subject to the deposits and acid overflow and will need to be replaced. Again, not a major expense, maybe $20, but look again. After I bought my son his '76 Firebird, we went out to replace the battery and tray and found that the corrosive gunk from the old battery had eaten through the inner fender metal that the tray was screwed into. Lousy inspection job and I paid for it with four hours of work just to get the tray housed so we could get the battery in.

And beware the guy who says: "Well, we can get a tray in. All we have to do is leave the battery at a slight angle . . ." When the battery is at a slight angle, the fluid inside the battery isn't in contact with the metal cells and you invite a dead cell.

Also note whether the car is one to three years old and already has a replacement battery. The original battery may have been defective but often when a motorist has trouble starting the car and doesn't know a lot about cars, he or she will mistakenly go out and buy a new battery. A replacement in a one or two year old car could mean that there are major electrical problems lurking that a fresh battery will cover up until the first sharp cold spell or a few weeks of severe heat and the drain from all the accessories will force the problem to surface.

Next check out the oil system in the car. Oil is the life-blood of a car and any indications of neglect here are a good sign that the car hasn't been too well cared for and that the life expectancy of the engine may not be what it should be.

Begin with the oil filter. Is it caked with grease and oil and assorted road grime that hints the filter has been on the car for many more miles than it should have been?

And what make is it? Personally, I like to see a name brand filter looking up at me rather than some off-brand filter I never heard of. An off-brand can mean one of two things: (1) the owner was more interested in buying a "cheap" filter than the one designed specifically for the car, or (2) it's hard to find filters for this car and you'll have to send off for them.

If it's a private party, ask the owner how often he changed the oil and filter. Most sellers trying to dispose of a car will be smart or shrewd enough to tell you that they were changed religiously. But the owner just might boast: "Why every 10,000 miles," or worse yet: "I don't remember." A clue to beware.

Of course, the owner just might come back and say: "Why, I change the oil every 2000 miles, use a Quaker State (or other brand name) filter and 10-W-30 oil which is what I advise you to do if you buy it."

Sounds good. It appears he paid attention to the car's upkeep. But go one step further and ask if he has any of the oil and filters left so that if you do buy the car, you can start with an adequate supply. If he shows you a case of oil and a couple of filters in his garage, chances are the car was cared for as he said. Not foolproof evidence, of course, but certainly a favorable sign.

Next pull the oil dipstick. Is the oil up to the proper level and in a fluid state or is there evidence that it is becoming solid? The latter would shoot holes in his promise of frequent care.

If the oil and filter pass the test, the next stop is the transmission dipstick. Here, the fluid should be pinkish or reddish. If it's brown or orange, it means it's burnt — and so is the transmission. Forget it, go look for another car.

If the color is right, ask when the fluid and the transmission filter were last changed. I hope you don't get a puzzled look in return.

On to the spark plugs. Take some out and examine them. If they are carbon fouled (see p. 142), don't be too surprised or concerned. It just means that the driver has done a lot of stop and go, short haul driving or is running on too rich an air/fuel mix. A fresh set of plugs, an adjustment of the air/fuel mix and running the car on the open highway to keep the deposits from forming again aren't going to cost you much in money.

But if you pull a plug and it is thick with oil deposits, you could be looking at evidence that the piston rings or cylinder walls are badly worn and now you are talking some big dollars. Make a note to have that checked by your mechanic.

Or the plugs might show signs of a brown or white crusty deposit on the side or the center electrodes and core nose. This is evidence that oil or gas additives are being burned during the normal combustion process. A buildup can cause the engine to misfire.

Stop and ask yourself the questions: Why did the owner use additives? Couldn't he get the car to start and stay running, or was he tossing additives in as a cheap way to avoid a tune, oil change or other maintenance? Ask the owner.

Actually you can get a clue about the condition of the car without even pulling any plugs. If it's a V-8 or V-6 with air conditioning, that means the plugs are difficult to get at for even the best do-it-yourselfer. Tell the owner you'd like to pull those two rear plugs to look at them. He just might say: "They're a devil to get to and I don't think we can," or "I've tried and just can't seem to get the wrench in."

You now have reason to suspect those two rear plugs came with the car off the showroom floor and have never been out of the car. The V-8 you'll be paying for will be working with the power of a V-6 which will affect your mileage, emissions and cold start capability. A valuable clue.

Next, check the cooling system. Is the radiator filled with coolant and the inside of the core clean? Or is it so loaded with floating debris or, worse, so caked with gunk that you can't see any of the core? Imagine what that poor engine has gone through.

Since the early '70s, most cars and especially the small four cylinder models have come equipped with plastic reservoir tanks—what the industry calls "burp tanks." This tank will have readings marked on the side indicating the proper coolant level for when the engine is cold and hot. If the engine is cold, check to see that the fluid level matches the proper mark. Later, after your test drive check to see that the coolant has reached the hot level mark as an assurance that it hasn't been left along the road somewhere.

The burp tank was added by the manufacturers because the coolant gets so hot in small and aluminum block engines and the overflow was spilling onto the engines and onto the owners' driveways. With the tank, the overflow goes into the tank. Also, motorists were scalding themselves when opening radiator tops to check the coolant. With the plastic tank you get a visible and far more safe reading on whether to add fluid and you can add the coolant simply by pouring it into the burp tank.

Open the tank and look inside. Do you see a layer of slime inside that indicates that the radiator has never been flushed and the slime has backed up into the tank? Is there solid matter in the tank? If there is, there is solid matter in the cooling system, too. Another sign of trouble ahead.

Give all belts and hoses a look and a squeeze or a touch. Heat kills rubber and you can expect hoses and belts to show signs of wear. Check the drive belt (also called the fan belt). Is it cracked or glazed and in need of replacement?

My concern would be uneven wear on one side of the belt or signs of unusual wear that may mean the pulleys aren't in line. I once owned a 1972 Vega and the belt had to be replaced regularly because the pulleys wouldn't line up. It's a low-cost item but it's a big aggravation when you have to carry a spare belt 12 months of the year and don't know if the belt is going to snap every time the temperature hits 20 degrees.

Belts and hoses are relatively low cost items to replace. But if you see signs that they all need replacing, point it out to the seller and ask him to come down on the price to match the cost of replacement. When you do replace hoses, flush and replace the coolant, too.

Give the engine compartment a good lookover for signs of leaks, caked oil or grease over a seam or for other signs of abnormal wear or abuse. If you see any, ask about them.

Don't close the hood until you've checked the air cleaner. The air cleaner is like the inside door panel: it's easy to spot neglect by checking it out. That's because replacing one is the easiest do-it-yourself service item you can perform. You can do it in five minutes with a cast on each arm.

Reach for the wing nut on what looks like a covered frying pan. Undo it, lift the top cover off and take out the cleaner. Is the cleaner element packed with dirt and dust? If it is, that dirt and dust has been getting into the fuel system. Are there oil marks on the filter or in that frying pan? That probably means that the engine has been so neglected that the positive crankcase ventilation (PCV) valve has never been changed and oil is blowing back into the cleaner element.

Close the hood. Depending on what you found underneath, you are now at a point where you may want to stop the inspection and head off to look for another car. If not—if you spotted nothing in the engine compartment that bothers you or arouses your suspicion—then you are ready to move on to inspect the glass and tires.

Most potential car buyers tend to overlook the glass when inspecting a car. That can be a mistake. Glass is expensive to replace and when it comes to the windshield, it's always a gamble that the replacement won't leak.

Check each window for pit marks, cracks or any unusual discoloration. Look closely at the windshield for signs that the black rubber moldings have been replaced, such as new rubber on a five year old car, or that a clear plastic silicone sealer might have been used to stop leaks.

The windows might also tell you something else. Are there decals on the window or the remains of glue indicating that there were? Decals will give you a quick insight into the owner and are a strong indication of the kind of travel the car has been put through.

A window full of travel decals representing points near and far means that the car has had a lot of miles put on it. A decal from a university could mean that the car was sitting up at school for nine months without an oil change.

If the decals are still on the window, look at them and draw your own conclusions. If they've been removed and there were a lot of them . . . well, that in itself says something. If there was just one, you might ask what it was. You can judge the answer on the basis of how you judge the person.

As for the tires, they often provide clues to present or future problems. And since a set can cost you up to $200 or more, the condition of the tires themselves is important.

Walk around and examine each one carefully. Do all four match? You may find the tires on the used car you are inspecting have mixed bias-belted with steel-belted. You'll know the difference in ride, handling and cornering, perhaps vibration or problems in steering. You'll want to switch to all four of the same construction.

It may be you will spot irregular wear on the outer ribs of the tires, indicating that the owner has been driving on underinflated tires. The wear is visible but you know that with proper inflation you will be able to get at least another year or more out of the treads. A bargaining point but not a reason to say "no" on the car.

But what if you see that the outer ribs of those tires have cup marks and the grooves are very deep and the tires are really beyond saving? Everything else about the car looks good and you can get a deal on tires for $150 a set and that's a price you're willing to pay for this car.

Think again. Cupping is a sign that indicates the shocks are gone. And if that's the case, you will need in addition to the tires a set of shocks which is a $50 to $60 item. That makes it $200 and that means it's bargaining time.

How about kicking the tires . . . aren't you supposed to do that? Only if you want a set of stubbed toes. That's the only thing that that advice ever accomplished.

OK, so now you've gone over the outside and the car's still in the running. Now it's time to go over the inside.

Begin by sitting in the driver's seat. Are the springs all right with no sag? Move the seat back and forth. Is there ease of movement? I once had a reader call to say he couldn't get the seat to slide back. After several minutes of trying he got out, looked underneath — and found a kid's shoe blocking the path.

What type of material are the seats made of? Vinyl is easier to clean than cloth but it will burn your butt in summer and freeze it in the winter. Inspect the seats. One small tear can be sewn but lots of them may mean covers. Another expense.

How about the carpeting? Is it in relatively good shape or is it worn and full of stains or gook? (Another sign that the car was probably used by a teenager or with small kids. Find out which.) Are you going to have to replace it? Another bargaining point.

If the vehicle you are looking at is a wagon and it doesn't have a carpeted cargo area, don't be discouraged. Buyers often are impressed that the cargo area the kids will play on is carpeted from the factory. The joy turns to sorrow when the first can of pop is spilled. You can't remove the carpet and clean it.

But if you find a wagon without the carpeting you can go to a carpet remnant store and pick one up cheaply. Then, when the kids spill or you've carried dirty or dusty cargo in back, you can remove the remnant, vacuum or shampoo it, and put it back again.

Now look at the odometer. Does the mileage match with what you've seen of the body and engine? Do all the numbers line up or is one out of line a shade, hinting of a rollback that took 25,000 miles away in one magic swoop?

Check the pedals. The brake, clutch (if it's a manual shift car) and accelerator pedals should show some signs of wear. But if the odometer reads 30,000 miles and the pedals are worn through to the metal, that's an indicator of at least several thousands more miles of usage.

Or does the odometer show 30,000 miles and the pedals are all new? Or is the accelerator pedal showing reasonable wear but the brake pedal down to the thread? Perhaps the owner did a lot of city driving which required stops and starts and use of the brake pedal. But perhaps also Hot Rod had to stand on the brakes a lot as his only means of stopping the machine from frequent 70 mph bursts.

How can you tell? Maybe there is no way but, then again, maybe there is. Are you buying from a private party? Is he wearing a T-shirt or hat that says "I Love Indy"? Or if you peek into his garage, is it filled with racing posters or, worse yet, ribbons and trophies he's won, perhaps using the same car he's showing you. At least grounds for suspicion.

Back to the interior. Check out every dial, knob, handle and button on the dash, console or arm rest. Do all the gauges work? Does the air conditioner send out cold air or mildly cool air? Does the heater send out hot air? Does the defroster send out a stream of air? Do the wipers wipe and the fluid squirter squirt?

Sound the horn. Check the lights inside and out, from the dash to the dome, the head to the tail. What about the reverse and brake lights? The turn indicators?

Turn on the radio. Do all the speakers work? Is there a cassette player? Try one. Is the owner a 60 year old man and all the push buttons on the radio are set to rock and roll? Someone else has been driving this car . . . someone younger with a heavier beat . . . and a heavier foot on the accelerator and brakes.

There is one other thing that you need to look for on a used car but as with buying a car for a teenager, this is so specialized that it requires a chapter all to itself. Which is exactly what it is going to get.

Using a checklist when inspecting a used car can be helpful. Here's one you might want to follow.

Inspection Checklist

Item	Condition 1. Good 2. Fair 3. Poor	Item	Condition 1. Good 2. Fair 3. Poor
Door panels (inside)	_____	Stop lights	_____
Ceiling	_____	Tail lights	_____
Upholstery	_____	Directional lights	_____
Dash	_____	Side marker lights	_____
Engine gauges or lights	_____	Instrument panel lights	_____
Carpets	_____	Dome lights	_____
Lighter/ashtray	_____	Engine compartment and trunk lights	_____
Trunk/spare/jack/wrench	_____		
Windshield	_____	Mirrors	_____
Windows/handles	_____	Doors (outside)	_____
Tires (same construction) / wear	_____	Fenders	_____
		Wipers	_____
Hub caps	_____	Window defrosters (front and rear)	_____
Moldings	_____		
Bumpers (signs of hitch)	_____	Horn	_____
Grill	_____	Hood and trunk release	_____
Headlights	_____	Fuel filler release	_____
Reverse lights	_____	Battery	_____
Parking lights	_____	Radiator	_____
Hazard lights	_____	Fluid levels (oil, coolant, transmission, brake, power steering)	_____
License plate lights	_____		

Item	Condition 1. Good 2. Fair 3. Poor	Item	Condition 1. Good 2. Fair 3. Poor
Filters (oil, trans., fuel line)	_____	Plugs (wear signs)	_____
Brakes	_____	Belts and hoses	_____
Seats (power or non-power and reclining)	_____	Undercarriage (rust, weld marks)	_____
Tilt or telescoping steering wheel	_____	Body panels (rust, repairs, replacements, fits to other body panels, signs of accident)	_____
Radio/speakers/antenna/ cassette	_____	Leaks (black for oil, pinkish trans. fluid, yellow green coolant)	_____
Clock	_____		
Heater and air conditioner	_____		
Exhaust system	_____	VIN (present and not altered)	_____
Converter	_____	Title (clear and in owner's name)	_____
Suspension (shocks, struts, springs)	_____		

THE INSPECTION
PART 2

Emissions control

If you live in a state which requires annual emissions testing you must be especially careful when purchasing a used car to ensure that the catalytic converter is in place—and that it is working. Buy a car without a converter and it's up to you to have one put on if you want to pass the emissions test. And a replacement converter can run $100 to $300.

To date, emissions test programs have been instituted in 31 states covering some 45 million vehicle owners. These programs typically require motorists to bring their cars in for an emissions inspection once a year. If the car doesn't meet state or federal standards, the motorist has to take the car back again—after paying for the work required to bring the car into emissions conformity.

If the car's catalytic converter is missing or if any of the other emission control devices that originally came with the car have been removed, tampered with or altered, the owner must replace them at his or her own cost in order to pass the annual test.

The converter looks somewhat like a muffler and is located under the car, roughly under the passengers' floorboard. Converters were made standard equipment on the majority of domestic cars and many light duty trucks in the 1975 model year when the government required the automakers to build vehicles that spewed out fewer pollutants than they had.

The converter contains chemically coated pellets or beads that treat exhaust gases as they pass through to reduce the volume of pollutants that will exit the exhaust pipe. Since the lead in gas coats the chemical materials inside the converters and renders them useless, the petroleum companies had to market lead free gas. And to ensure that catalytic converter-equipped cars used only lead free, the petroleum companies had to install a narrow nozzle on the lead free pump and the auto makers had to place a narrow restrictor opening in the cars' fuel tanks so that the wider leaded regular nozzle wouldn't fit.

The public at first didn't take kindly to converters or lead free fuel. Converters, many falsely felt, robbed engine power. Lead free fuel, many truly found, cost more than leaded regular and was more costly to use. So, thinking they'd improve performance, many car owners simply removed the converter. Others, to save on the weekly fuel bill, doctored the car's inlet restrictor so that they could pump leaded fuel into a lead free tank.

In 1979, following the Arab oil embargo when gasoline prices skyrocketed and supplies allegedly dried up, the cost of lead free fuel rose astronomically and more motorists joined the parade to remove emissions hardware so that they could use the cheaper leaded regular fuel.

Many of those cars are now on the used car market, so *beware*. The Motor Vehicle Manufacturers Association, citing a Survey Data Research, Inc. report, said in 1976 it was found that 18.7 percent, or almost one of every five new car buyers removed the converters from their new cars. By 1984, when gas prices stabilized, only 1.4 percent of car owners removed the converters.

When shopping for a used car from a dealer, have him show you the converter to ensure it's there and have him sign an affidavit that all federally mandated pollution control hardware is on the car and if any items have been removed they will be restored at his expense. It is illegal for a dealer to remove the hardware. But it is not illegal for him to sell a car that he took in trade without that hardware.

It would also be to your benefit for him to run an emissions test on the car to ensure it passes state requirements *before* you buy it. But, he'll argue, that will cost *you* money. You argue back that if the car doesn't pass the test it still will cost *you* money, so have it done. Besides, if he's asking $5000 for the car and a converter is going to cost you $300 to add later, you're actually going to have to pay $5300 for the car. If he won't add the converter, he better reduce the asking price to take into account the cost you'll soon incur in having to put one on.

If you are purchasing from a private party, relative or neighbor, it gets a bit touchy, but again make sure the converter is on the car. If not, tell the seller: "Okay, you want $5000 for the car but it will cost me $300 to add a converter, so I'll give you $4700."

It's the same as finding out the car needs a $300 brake job. You use your upcoming expenses to negotiate a better price.

Finally, just because a converter is on the car you have no way of knowing if that converter is actually working. That's one reason to have the dealer, who has the test equipment, run a test on his own. But also check the gas tank inlet restrictor on your own. As I said, many motorists doctored the inlet in order to allow a leaded regular nozzle to fit. If the inlet has been doctored no doubt the converter has had more than the couple of tanks of leaded fuel needed to render the unit useless. It again means you'll have to replace the converter.

Keep in mind that under federal law each auto maker must warrant its cars' emissions systems for five years or 50,000 miles. That means the components in the system must be free of defects for that time and mileage period or replacements have to be put in at the factory's expense.

If you've used leaded gas in your car designed for lead free only and the converter doesn't work, you pay for any replacement. If you've used lead free only and the converter no longer functions after 35,000 miles, the factory has to put a new one on. Keep that in mind if you fail a state emissions test. Before running out and having a tune at your expense, first ensure the problem isn't an emissions-related warranty item.

I had a reader complain that her car failed the test, found it was the converter, had the dealer replace it free under the federal warranty—and then charge her a $45 "inspection fee." If this happens to you, call your local Environmental Protection Agency office at once. Also call the zone office of the company that made your car and contact the manufacturer back in Detroit.

If the car is just what you want and the seller will either replace the emissions hardware or reduce the price to offset your expense in having to add it, go ahead and buy.

And one last note: diesel engine cars don't have converters and in most states aren't required to take the emissions test. But if you think that's a good reason for buying one you haven't been paying attention.

BRINGING IN THE PROS

Your mechanic and your insurance agent

OK, you've gone over the car outside and in and you like what you see. The car seems sound, the seller seems honest and the price is about what you intended to spend.

Time to climb out and close the deal, right?

Wrong. Time to call on the pros.

Before you buy any car, there are two professionals you ought to bring into the process. One, as most people know, is your mechanic. The other, as most people *don't* know, is your insurance agent.

Your insurance agent?

Yep. It could save you a lot of money and aggravation. It could even save you your car.

Most often we go through the entire process of shopping for and checking out the used car, buy it, drive it home and *then* call our insurance agent and tell him or her that we just bought a 1984 Spritzel and need insurance.

"That's interesting," he says. "Based on the accident repair records of Spritzels, our company tags all owners with a 15 percent premium on those cars. You were paying $300 every six months on your old car. You will now be paying $450."

"Now you tell me," you moan.

"Now you tell *me*," he replies.

He's right. The time to tell him or her is *before* you buy

the car, not after. Do it after your own initial inspection but before you call your mechanic to give the car his professional once-over. What the agent tells you may make the mechanic's trip unnecessary.

He may tell you that the car you are considering buying typically costs a lot more to repair than other models and therefore carries a premium surcharge. With some insurance companies that can mean up to a 30 percent penalty.

He may say not to worry about making repairs because the car is stolen so frequently that you won't have time to even perform an oil change. That certainly changes the picture.

He may say that he insures others with the same limited volume import and has found that they simply can't find replacement parts or have to wait weeks while the car is laid up in the shop. Or he may say that he insures others with the same limited volume import and has found two local independent mechanics who do the work, charge fair rates and are so competent that the owners don't have to go back two or three times to get the job done right.

He might even say that the accident and repair record on the car is so good that you'll get a 15 to 30 percent *reduction*, which is something to factor into your decision when the owner won't come down another $50. Go ahead and buy — you'll get the money back in lower insurance premiums.

The other pro you should call on, of course, is your mechanic. Most people know this. What most people *don't* know is *how* to do it.

Your mechanic *is* a pro — and if he or she is any good — a *busy* pro. That means two things: (1) that you should expect to pay him for his or her time and expertise, and (2) you shouldn't waste his or her time.

Finding a mechanic you can trust has become virtually a necessity of modern life. For most of us, it's a matter of trial and error. But when we find one, we hang onto him for our car's dear life.

Mine's name is Bill. I won't give you his last name because I want to protect him. For me.

But this is the kind of a mechanic Bill is.

Remember that Firebird I bought my son? The one with the corroded battery tray? Well, it turned out when we got it home that it also leaked transmission fluid. (That car served its purpose well. It taught my son not to go with glitz.)

I replaced the seals and refilled the fluid. The next day there was another puddle of fluid beneath the car. Seals replaced and fluid refilled again. Next day another puddle. A frantic call to Bill, my independent mechanic without peer.

"Go to a dealer," said Bill, "and make sure you have the right transmission dipstick for the car. Sometimes on an old car when a guy tries to restore it, the stick will be bent and he'll just grab one from any similar make and think he's got the right one."

I went to the dealer. Sure enough, the dipstick that belonged in the car was about four inches longer than the one in the car now. That meant that each time fluid was put in, too much was added in order to reach the "full" line on the short stick. The fluid heated up and with the pressure it had nowhere to go — except through the seal and onto the driveway. The new stick was put in and the puddles on the driveway ended.

Another time: A battery problem had me stumped. The temperature was well below zero for three days running and for three consecutive days the battery wouldn't turn the car over. Each time I went to the store and got a different battery. It worked fine during the day but the next morning — again, no life. I called Bill. Bill suggested: "Open the trunk and see if the light switch is broken in the 'on' position." It was. No more trips to the store for batteries. And no trip to the shop with parts and labor included.

That's the kind of mechanic I'm talking about. The kind you want to cherish and not waste his time — if for no other reason than you want him to respond when you call.

Wasting his time is calling him out to see every car you look at. That's not the way to do it. It's also a waste of time to call him up after reading the classifieds and rattle off the names of 18 cars you spotted and not only ask him about the cars but whether the price sounds good.

The right way to do it is to place your call at the very beginning of the process, before you start looking, and tell the mechanic that you are in the market for a used car and you would like to be able to call on him to evaluate a car or two after you've made your own inspection.

Call him first thing in the morning before he gets busy or leave a message telling him what the call is about and ask him to call you back at his convenience. Or drop around the shop and wait until he has a break and can give you a few minutes.

It may be that your mechanic can give you some help without even leaving his shop. He or she deals in cars every day. Nobody is likely to know more about track records, unique problems that don't get a lot of publicity, trends that develop such as a particular part in a particular line of cars that keeps malfunctioning. He or she is the one who keeps a mental file of cars brought in that seldom need more than a tune-up or filter change and are back out on the road again.

Got your heart set on a '78 Spritzel? Ask your mechanic about it.

"I got six of them lined up out back and four more coming in tomorrow," he might say. "The brakes last a week, the calipers have the strength of tissue paper."

Or he might observe: "A '78 Spritzel, huh? Good choice. I haven't had to make a major repair on one of those in three years. Runs forever."

He might even know of a customer who's looking to sell one before buying another car. And he'll know whether it is or isn't a good buy. He should. He has been taking care of it.

The first call or visit is also the time to make your financial arrangements. How much will it be? That depends on the mechanic and where you live but it should range between $20 to $50. Sound like a lot? It might — but $5000 is a lot more and spending one percent of that to ensure it's a good investment does not seem to me to be excessive. To me, that seems wise.

After that, you should not call your mechanic again until you've done your shopping — and inspecting — and you've found one or two cars that you think might be worth buying.

Don't call him about the car that you looked at and the entire right side has been restored . . . and the radiator has solid matter inside that you can pick out with your fingers . . . and the oil on the dipstick is crusty . . . or the rocker panels are loaded with plastic body filler. You don't need him to tell you not to buy that car. You already know you don't want it.

Call him when you need him. When you've found the car you think you want.

He'll appreciate that. And he'll respond when you call.

THE TEST DRIVE

You've given the car a visual inspection. It appears to have been well cared for and none of the major danger signs have reached out and grabbed you. It is now time to call for your mechanic and go for a test drive.

New or used, taking a test drive of the car you are thinking about buying is a must. Yet this is one area in which most shoppers fall short of exercising their rights or fail to take advantage of the expertise that is available to them. People about to spend $5000 or more on a car may spend less than five minutes checking out the car they hope to live with for the next five years.

Most shoppers treat the test drive like a person buying a new pair of shoes. They have the salesperson bring out the shoes in the color and size they want, put on the left shoe, stand up, walk three paces and say: "I'll take them." Only when they get home and put on both shoes do they realize that one doesn't fit or the leather squeaks.

It has always amazed me in my years as an auto editor that people will call or write to complain about a car for problems that they would have avoided if they had paid attention during the test drive — or had taken the test drive in the first place. In fact, let's not call it a test drive. Let's call it a learning experience because that's what it is — you're learning about the car.

If you have a knowledge of the basic mechanics of a car, then you are probably qualified to perform the test drive on your own. But if not, and most persons do not, this is the time, as I've suggested, to get your mechanic involved. He will know what the strange sounds mean, the fluid leaks warn of, what the telltale puffs of smoke coming out of the exhaust indicate about the inner workings of the car.

(Incidentally, if for some reason you don't know a mechanic — you've just moved, perhaps, or you've never owned a car before — you can try finding one through friends or ask a friend who is knowledgeable about cars to go along for the test drive. It's better than going alone. The risk, as always, is that if the friend misses something that later costs you a lot of money to repair, the relationship also might need some repair.)

The first thing the mechanic probably will do is conduct his or her own visual inspection. That's all right — he may spot something you overlooked or didn't understand. He may even spot something good the owner did to the car, like adding one of those do-it-yourself radiator flush kits, that hints at a concerned owner.

Once he's finished going over the outside, he'll give you the nod and tell you it's time to move inside for the test drive.

If the car is to be the family vehicle take the family for the test drive. Whoever is going to use the car should be in the car. Often parents will find a sitter for the kids when they go car shopping to keep them out of their hair. A big mistake. A combination of childhood innocence as well as an attempt to help Mom and Dad in a big purchase decision often brings out things that adults never would have thought of.

To kids, the back seat is home and if they find things they can't live with, they'll tell you. In fact, before you leave, tell the kids their job is to check out the back seat. They'll probably go overboard to find flaws, but all you need is for them to find *one* big one and it will have been worth bringing them along.

I pause here while speaking of kids to point up a common misconception among parents of little ones. The belief is that when you have young kids, you should own a four-door car. The extra doors make it easier for the kids to climb in and out without disturbing Mom or Dad up front. While lots of today's newer cars have childproof locks in back to protect the kids while you're driving on long trips, you may not have that choice on a used car.

Having brought up three children and having driven hundreds of different cars over the last 16 years, it's been my experience that parents can travel far more contentedly and concentrate better on the driving if they are in a two-door car with the kids in back. Then you *know* they aren't going to accidently open a rear door—child-proof or not. Sure it means fighting those seats every time they get in or out but it means far greater peace of mind. The four-door is much more convenient—and sensible—when you've reached the age that the rear seat passengers are fellow adults.

With everyone on board, do you find the car to be roomy and comfortable . . . or is it cramped? Power seats are nice but usually the power controls are in the arm rest and rob you of some arm room. The crank type window isn't as attractive and requires some work to operate but if you need one inch more of room and the car has power windows, you aren't going to get it.

With everyone inside, check the visuals. Can you see out the front, side and rear with no trouble? (That's another reason you bring the family along.) A blind spot on the used car lot or in the neighbor's driveway is going to be a blind spot on the open road, too. (The 1979–1985 Buick Riviera, Olds Toronado and Cadillac Eldorado are notorious for a rear quarter pillar blind spot that makes passing or a lane change a true adventure.)

Can you see and reach all the controls? How about all the other people who are going to be driving the car? Can they all see and reach, too? How about the teen who is going to drive? Keep in mind that the teen won't admit a problem if he or she is in love with the make, model or color so you may have to put them behind the wheel and say: "OK, now turn on the air." If he or she has to bend below the dash level to do it, forget it.

As a rule, when you can see and reach controls with your right hand, you have no problem. When you have to start probing with your left and search under the dash or along the console, look out.

The Dodge Omni and Plymouth Horizon feature air and heater controls along the bottom of the dash to the left of the driver which are difficult to see, much less use. And wait until the 1986½ Jeep Wrangler appears on used car lots. Its ashtray and cigarette lighter are *under* the steering wheel. Yes, *under*.

As I pointed out before, the car industry has undergone a downsizing program to make cars lighter and smaller and thus more fuel efficient. People, on the other hand, haven't gotten smaller and the laws of science say that at some point the two opposed factors have to meet. I find the meeting is at the stomach right at the bottom of the steering wheel. So if you are looking at a small car, check to see if it has tilt wheel. Without tilt wheel, some of the subcompact cars fit like a $29 suit. If more than one person is going to be driving the car, test them all behind the wheel.

If the car you are looking at is a wagon and you have kids, let them at it. Can they operate the seats to put them down and get them back up? If you have teens, or even pre-teens, you'll have to turn the radio up full blast for them to hear it unless it has speakers in the back.

Once you've tried the car for fit, it's time to turn on the key and take a ride. And I'm not talking about four left or right turns around the block at the dealership or at the house of the private party with the car for sale. I'm talking about testing the whole car, the vehicle that you may soon be spending $5000 or so on.

If you are at a dealership, you don't want the salesman along if you can avoid it. The salesman tags along for three reasons: (1) to use the time to try to talk you into the sale, (2) to ensure that you only make four right or left hand turns and not put the car through its paces, or (3) to make sure you bring the car back.

Cars have been known to disappear on test drives. In fact, there isn't a dealer in the business who can't run off one or two favorite stories about the nice elderly gentleman or the sweet young couple with kids who left for a test drive and never were seen or heard from again.

The private party will probably also want to accompany you on the drive. I wouldn't blame the owner because, for all he knows, you could be a thief.

Readers have reported to me instances in which the dealer has refused to permit a test drive. In some cases the dealer claimed he wasn't insured for test drives. Tell him to call an insurance agent and get temporary coverage so you can drive it and you'll return when that formality is done. Or you can say: "Well, Joe Jones down the street lets me drive cars. I'll go there."

If you are told the car can't be tested, your best bet is to simply pile back into the car you came in and take off because you're dealing with a flim flam man. It just could be that the car has so many obvious mechanical flaws that even a drive down the block would bring them all out.

In some cases today, usually with hard-to-get new Japanese imports, some dealers have been known to refuse a test drive until the potential customer has made a down payment, which the customer later learns is non-refundable. Total gall. The intent is to handcuff a buyer. You don't like the car you drive, fine. As long as I have your deposit, let's look at another. And another. You'll keep looking until you can use that deposit on a car he has in stock.

If the salesperson or private owner does want to tag along, just ask them to buckle up their seat belt and please sit back and be quiet. If you have any questions, you'll ask them but for the next several miles you'll be busy.

You also don't want the radio blaring—or even on. You want to keep your ears tuned for unusual squeaks, rattles or sounds that you wouldn't normally expect in a car you might buy—sounds such as metal rubbing against metal from the brakes, a clunk from underneath when the transmission shifts gears, or other clues that could indicate the need for a major cash outlay if you were to buy the car.

Do your test drive on both city streets and open highway to test various aspects of the car under different conditions. Is acceleration from a stoplight smooth, quick and quiet or does the car lumber out of the blocks? On the highway how soon after you hit the pedal does the car react? Do you have the power to merge, climb hills, pass another car?

How good are the brakes? Does the car stop quickly and in a straight line or does the pedal go nearly to the floor and the car pull sharply to one side? Do you hear any metal against metal squeaks when the brakes are applied?

How about the manual transmission—does it shift smoothly or is it balky and hesitant? Usually, the Japanese models have it all over the U.S. makes for smooth, quiet manual transmissions. GM manuals generally are the worst among domestics which is why GM is having Getrag of West Germany design a five-speed for use in GM cars. The best GM manual is in the Chevy Nova subcompact but that's because the Nova is the joint venture car with Toyota and Toyota supplied the engine and transmission.

If it is an automatic shift, how quietly and smoothly does it shift? Does it balk in each gear or, worse yet, clunk when downshifting?

In the late '60s and early '70s there was a common belief that no automatic could hold a candle to a manual in terms of off-the-line performance and fuel economy. Automatics were noisy, sluggish and as a rule got from two to six mpg less than a manual-equipped car. That was true in the '60s and

'70s. It is not true in the '80s. Refinements — especially the addition of computer controls — has made the automatic shift as good as the manual today.

Your test drive should also take you around some sharp corners and turns. Does the car hug the road or wander off to one side because the suspension is badly worn? Does the rear end bottom out on a bump? Does the rear or front end keep moving up and down several times after going over a bump because the shocks and/or springs (expensive) are worn?

On the way back to the dealer's lot or owner's driveway step on the accelerator pedal and then back off. Look for traces of smoke coming out of the exhaust. Bluish smoke would mean the engine is burning oil. Black indicates that the air/fuel mix is too rich and an adjustment is needed. Gray smoke denotes transmission fluid and the possibility of major and costly repairs.

When you arrive at the lot or driveway wait a few minutes and then get out and look under the car. Check for fluid leaks. Is the pavement dotted with black spots from oil, pink ones from transmission fluid or greenish-yellow ones from loss of coolant? Wait a few minutes more and see if the spots become puddles.

Your mechanic probably will have known in seconds if anything was amiss and may want to give it another visual check when the drive is over. If you do the test alone, you will have to depend on whatever clues you have found.

You may have found enough to get in your car, thank the salesman or seller, and go home. Or you may want to reach for your checkbook if your mechanic is grinning from ear to ear. If you've tested the car on your own, you may want to consider another test drive at night. You may find a problem you couldn't spot in the daylight, like poor visibility or a weak electrical system. You may also want to try the car after a rain to double-check the suspension, ride and handling in adverse conditions. I've driven Saabs on dry roads and have been impressed. I've driven Saabs on wet roads and been frightened to death because of the poor road-holding ability.

Buy now or wait? I suggest you wait. Go sit down with the mechanic if you brought one or go home and sit and reflect if you did the test drive on your own.

Why?

Read the next chapter.

THE DECISION

Adding it all up

The test drive is over and you and your mechanic find a nice quiet spot to discuss just what you've found wrong with the machine.

Personally, you thought the car looked great, rode and handled very well, and the price is certainly within the budget you laid down. Gushing with excitement you ask your mechanic: "Well, what did you think?"

He looks you straight in the eye and responds:

"All the belts and hoses need to be replaced, the battery is on its last leg, the shocks are gone, and chances are you'll need a complete brake job."

Your jaw drops and your cheeks take on an ashen hue. Just when you thought you had found a car you could depend on, your mechanic puts a kibash on the deal.

"What's the matter?" the mechanic says after noting you've reached for the glycerin pills to get the heart pumping again.

"You just told me the car isn't worth buying," you reply.

"Never said that," the mechanic answers. "In fact, if I were you, I'd go back with your checkbook. You've found a pretty good car."

The problem with shopping for a used car is that we tend to expect total perfection and if we don't get it, we give up on the machine. Remember, the reason it's called a *used* car is the fact someone else has owned it and put miles on it. Seldom does a person decide to get rid of his current car and trade it in on a new one when his current car is running like a top or looking showroom new.

There still are some people today who trade a car in every one or two years just to have a fresh new model sitting in the driveway. But those people are few and far between.

The reason you have a mechanic check the car out in the first place is to find all the problems that exist and *then* come up with an estimate on what it will take to make the repairs if the repairs are indeed worth making.

Let's take the example above.

I'm going to use ballpark figures obtained from Chicago-area repair facilities. The costs will vary by the type and size of car you're looking at and the region of the country you live in. Chicago prices generally will be a shade higher than Georgia, for example. But with that qualification, let's look at the car more closely.

"All the belts and hoses need to be replaced" — Replacing all the belts could run $60 to $70 and all the hoses $100 to $125. It's gotten costlier in recent years with all the heat those hoses and belts have to withstand as well as the fact that more automakers have gone to complex one-piece serpentine belts which are much more difficult to replace.

"The battery is on its last leg" — Costs have risen substantially because new chemical compositions are used in so-called maintenance-free batteries. But batteries frequently are loss leaders or sale items at local parts stores or mass merchandisers such as Sears. You could pay $40 to $100 installed.

"The shocks are gone" — Here again you'll find shocks on sale almost as often as you find a weekend special on 8-packs of soda pop at the local grocery. You might pay $50 to $60 installed, though you could pay even less than that depending on the specific sale.

"Chances are you'll need a complete brake job" — This

isn't a low cost item if you need pads, shoes and lining. You might get away with a $129 special or go up to $180 to $200.

Let's look at all the work to be done on this particular car: $160 for all new belts and hoses, $50 for a battery, $50 for a set of shocks, and $200 for brakes. That's $460 total.

And that's not bad, unless the car you are looking at is selling for $500. But $460 would be a small price to pay for a $5000 car — less than 10 percent of the asking price.

"I had a customer come in and we figured he needed $1500 worth of work for a $5000 car and I told him to forget it," a mechanic told me who regularly performs inspections and test drives for motorists. "But I had another guy come in and we figured he needed $800 to $900 work on a $5000 car and I told him to buy it."

Be prepared, in other words, to put some money into the car and don't be discouraged and give up on what could be a very good car after a few dollars and repairs have been made.

Let's go over some of the other items mentioned in the inspection and test drive. You notice uneven tire wear and after a water test determine the car needs a front end alignment. Roughly $20 to $30. If the tires are badly worn $30 to $80 each to replace.

The air conditioner is pouring out a lukewarm blast because the condenser needs replacing? $45 to $70.

You open the radiator cap and the matter inside was solid? A flush will run about $30, a new radiator about $150.

The exhaust system is playing in baritone and needs to be replaced. With so many franchise exhaust stores around it's not hard to find a weekend special. Depending on the car and the special being offered, you could pay $50 to $150.

The exhaust smoke is blue, meaning you're burning oil, and the plugs you pull are oil fouled. Now you may be talking a new or rebuilt engine and lots of money, easily $1000 and perhaps $2000.

How about the gray smoke out the exhaust and the fact the transmission was slipping or clunking when you shifted?

A rebuilt transmission could mean $275 to $500 in a rear drive car, $700 to $1200 in a front drive car. If it's a front drive, your mechanic might not even want to bother because getting to the trans means removing the engine first.

You notice a metal on metal contact noise when applying the brakes. Usually it means a pad or shoe, but on a front-wheel drive car it could mean a bearing is shot and that's $90 per side.

Or, you notice when applying the brakes the car pulled sharply to one side. Could be the caliper or piston is sticking ($80 to $100) or the master cylinder is shot ($60 to $110).

Valves? Very expensive, maybe $450 for the average car, double that on a sports or specialty model.

Struts? When Detroit started downsizing its cars and making them smaller to make them more fuel efficient, it found MacPherson struts helped a great deal. Rather than separate shock and spring, the strut combines the two in one. Space was saved but at a price. While shocks could run $50, struts could cost you $140 to $200 for a set.

After sitting down with the mechanic and getting an estimate on what needs to be done and what it will cost you, you have several choices in front of you.

If the cost of repairs is prohibitive, simply take a pass and look elsewhere. You don't have to buy the first car you see.

Or maybe you don't mind the cost of the repairs, but the fact that the owner left so many minor things undone really bothers you. If he neglected to invest $50 in a battery and $45 in an air conditioner compressor, he put up with hard starts and hot air in August, and you have to wonder what major problems he put up with. If simple maintenance was neglected, what about the serious stuff? Take a pass.

Or, you tally up the work to be done and determine it's a small price to pay for the car and realize that with the repairs made, you can expect three years or more of operation out of the car.

Be prepared to buy it — but again, don't rush.

Have your mechanic make out a list of the problems he found in the car and estimates of what the repairs will cost. Take the sheet to the seller and try to negotiate. He's been asking $5000 for the car? Show him you will have to spend $460 in repairs and try to get him to take $460 off the asking price or at least split the difference and knock $230 off. Any savings you negotiate will help defray your costs.

He has the choice to tell you to get lost or to make a counter offer. He may say: "I know a guy who can do that work cheaper than your man," and rather than come down $460, he'll offer to come down $200. Fine, you've helped yourself.

Or he may say: "I know a guy who can do that work cheaper than your man, and I'll have it done and deduct that cost from the asking price." Not bad if you can get him to agree to have your man look over the other's work one more time. If he refuses, I'd look elsewhere.

But he may say: "I just had my man look over the car and that work really isn't needed." Tell him his mechanic is in need of glasses and take a walk. You're either dealing with a fool, or he thinks you are.

Suppose you like the car mechanically and structurally — everything is sound or can be made so — but you aren't too crazy about the looks. It's an economy car, for instance, and you'd like something with a little more glitz. Or you don't like the way the inside looks.

No problem. With a few tricks and gimmicks you can make that little economy sedan or coupe look quite attractive and transform it into a pleasant-looking vehicle you'd be happy to own.

The secret is to find a good detailer. Detailers are the people who clean up cars taken in trade at a dealership and make them presentable for the used car lot. As one detailer once described his job to me, it is "to perform a cosmetic prep on the trade-in before it goes on the used car lot." That's a good description.

You can get one to do the same thing for you. Check the Yellow Pages or talk to a dealer or two. And of course check out their reputations. I once asked a dealer about a detailer I had interviewed. The dealer didn't think too highly of the detailer. "He charges too much," the dealer complained. I interpreted that as a compliment.

If you're handy and want to save some money, there are some things you can do to the car yourself. You might want to put on some of the add-on moldings that now come in a variety of colors. Or add a plastic rear deck lid spoiler or a louvered rear window cover. Or put on some decorative wheel covers. With a little effort and a little money, you can convert a pauper into a prince.

But if you want something done that really requires the professional touch—say an orange pinstripe from front to rear on that sterile yellow two-door Chevette you just bought your son—take it to the detailer. He'll not only dress up the outside, most of them will also dress up the inside. They will wash, wax and buff; vacuum and shampoo carpets (including the trunk) and upholstery; clean the chrome and windows; and even degrease the engine.

In short, a good detailer will make your car look like the last time it went through the car wash it was the Fountain of Youth. And that's a good investment.

THE PURCHASE

You've made your decision and it's time to go back and complete the sale. If it's with a private party don't hand over dollar one until you check that the car has a VIN number on the dash and therefore isn't a stolen piece of merchandise that you'd lose the first time a sharp cop spots you in it. If the VIN plate is there but looks as if its been doctored or altered in any way, forget it. All you can do is lose.

Ask for and examine the title. It must be in the owner's name and not carry any warnings such as taxi, rental, repo, salvage or the like. If the title is made out in the names of both Mr. and Mrs. Jones make sure that both Mr. and Mrs. Jones sign it—and *in your presence*. It could be that Mrs. Jones is in the process of divorcing Mr. Jones and he's trying to get rid of community property. Don't make the mistake of getting yourself tangled up in that.

No title? No sale!

If you're told that the title is being mailed from a former residence and will be forwarded to you when it arrives, you reply that you'll complete the sale when the title does arrive.

Make out a bill of sale, have both parties sign and make sure that each of you gets one. Also ask for an odometer statement that the mileage shown is accurate.

Don't pay by check or cash. Have a cashier's check made out by your local bank. And have it made out to yourself.

That way, if something goes wrong at the last moment, you will have no problem redepositing the money or cashing the check. If the check is made out to the seller, you could have some problems.

If you are purchasing the car from a dealer, follow the same procedures — but add some others.

Many Americans have been lured into the new car market as a result of discount financing programs — offers as low as zero interest rates from the automaker's own financing subsidiaries on two year loans. But those low rates are on *new* cars. *Used* cars carry much higher financing interest rates.

Before signing any paper, read it thoroughly to ensure that if the salesman said you're paying 11 percent, you indeed are paying 11 percent and not 22 percent. I've come across cases in Chicago in which the buyer *thought* he was paying one rate only to find out after he got home and started reading the papers that he was paying 33 percent on a used car loan. Yes, 33 percent. One dealer told me a customer came into his store wanting to trade in a used car on a new one. The dealer looked at the dealer decal on the rear fender and quickly realized no sale could be made. The dealer who sold the customer the car he was driving was notorious for 20 to 30 percent interest rates on used cars. Sure enough, the car was worth $2000 in trade but the man owed $4000 on the car and was locked into a 22 percent loan rate with the other dealer.

You want to see in writing what the rate is you're paying on an annualized percentage basis or APR.

If the salesman says he'll fill in the loan rate ("Hey, I said 11 percent, didn't I? Trust me.") and mail you a copy, don't stand for it. That paper could come back marked 22 percent and if you've signed, it's the old his-word-against-yours routine. Guess who'll win?

No money should be paid until all papers are completely filled out and you take the time to slip off in a quiet corner and read all of them. Any questions, ask. If you don't get answers, you don't pay. And remember, you want the answers in writing. No verbal promises. And no blanks on any paperwork and copies of all documents *now*. Don't go for any: "We'll mail it."

A few other words of warning. Don't be surprised if the dealer tries to sell you a service contract. As we mentioned in the chapter on the used car window sticker label, the car sold at the dealership must inform you if the car is sold "As Is" or with a limited warranty as well as note if a service contract is available and at what price.

If there is a warranty, get a copy. Again, don't accept: "We'll mail it." Chances are you're paying extra for a service contract and really not getting one, rather the dealer is gambling against a comeback. Or if you saw the contract and found out all the items excluded, you wouldn't touch the policy. Then, too, you may find at close examination that the service contract is an insurance policy with good old "Fly-By-Night Indemnity" that's only good if Fly-By-Night stays in business as long as your car does. And without a copy of the contract you won't know that one of the requirements is that all service and maintenance be performed at the dealer on a religious time and mileage schedule or the contract is voided, or that each service visit you make requires you pay a $25 or $50 deductible. Worse yet, you wouldn't see the fine print noting the contract is voided for owner abuse and abuse is defined as any time you drive the car.

If the car has been checked out thoroughly by your mechanic and if you take good care of your own car and regularly change fluids and filters, the odds are you won't experience major problems and can save money by taking a pass on the service contract.

Won't you make out if the transmission goes out in one year? Sure—providing it's not considered a case of owner abuse and the dealer refuses to reimburse because you failed to keep records and receipts to show you had the necessary maintenance performed. And what if old Fly-By-Night dissolved and is now old Ship-by-Day?

You also may be confronted with a series of gimmick add-ons to the bill of sale that you should be wary of.

These phony fees first popped up in the form of the documentation fee or documentary fee in the late '70s, brought on by a sharp decline in U.S. car sales. Dealers wanting to profit at any cost during hard times added a $7 to $25 DOC fee on the bill of sale. Since then the types of fees have grown in number, ingenuity and amount.

What's the documentary fee cover? Good question. Dealers that charge such a fee argue it's for paperwork. Those who don't charge the fee admit it's no more than a clever ruse to make a few more dollars off a customer who thinks it's a legitimate charge for a legitimate service when in fact it's a pocket profit.

The fees look official, especially to a buyer who hasn't purchased a car in five or more years, since they are printed on the bill of sale right along with such common and ethical charges as state and local taxes, license and title transfer.

In addition to a DOC fee you may spot a delivery and handling or D&H fee. This fee sounds as if the dealer did something to the car to make it ready for sale, like fix anything that was broken. Truth be known it may cover the dealer's luncheon tips.

Dealers, if nothing else, have ingenuity, as proven by the one who advertised that his cars carried a VHF. While sounding like the dealer was throwing in a TV when you bought the car, VHF stood for Vehicle Handling Fee, which reportedly covered anything the dealer did to the car to get it ready to deliver to the customer. In other words, he washed it.

Some dealers have been known to tack on a fee for cleaning the car before putting it on the used car lot. One dealer called this a Maintenance Fee.

Regardless of the name, the salesman or dealer typically will stumble in coming up with an explanation, and then it will be feeble. As I said, when you spot fees such as these printed on the bill of sale in type a slightly different size or boldness than the rest of the printing you can suspect it was a dealer add-on. You have the right to negotiate the charge off the bill of sale because the dealer arbitrarily negotiated it onto the paper. Refuse to pay. If he then refuses to sell you the car . . . walk out. If he plays games on the sale, he'll play games on service and warranty work too.

Some dealers, either in addition to the phony fees or in lieu of them, will try and talk you into extra cost options that line their coffers at your expense.

Rustproofing, for example, is a ridiculous option on today's new cars because the auto makers use so much galvanized metal and zincrometal in their body panels. They also add plastic nick- and scratch-preventing coatings along lower body panels, wheel wells and rocker panels.

But a dealer may try to talk you into rustproofing your used car. At best you'll be applying the rustproofing goop over metal parts already exposed to water and moisture.

Then there's the exterior body lusterizers that promise to keep the finish showroom new forever. Problem is the lusterizer warranty requires that *you* apply another coat of the stuff every six months. Another problem arises when you find the coating is starting to get chalky and you can't remove the film *unless you apply another coat of the substance*. Stay away.

Or you may find the salesman trying to talk you into a fabric treatment to keep all the spills from staining your seats. A can of Scotch Guard will do the same if you spray it on yourself, except that you'll pay $5 for a can of the stuff at the store but the dealer will apply fabric treatment at his place for $100 or more.

There's a pretty good rule of thumb to follow here: if the dealer or the salesperson is pushing it, stay away from it. It's probably something you neither need nor want. Or you can get it cheaper elsewhere.

CAR CARE
PART 1

Introduction

You are now the owner of a good used car that should last you several years — given the proper care. But giving proper care is what too many car owners don't do — to their ultimate grief.

In talking to car owners over the years, I have found that the ones who still enjoy their cars after 100,000 miles are the same ones who spend a few dollars each year on upkeep and service.

When I review a car, say it's lousy and advise readers against buying it, I'll always get a call the next day from someone who owns a three, four or five year old version of it.

"Listen," the caller will say, "I have that same car, it's five years old and I got 80,000 miles on it and it's still humming."

My first question to the caller always is: "And how often do you change the oil and filter?"

The answer invariably is: "Every 1500 to 2000 miles — both oil and filter."

And that, friends, is why the car has 80,000 miles on it and probably will top 100,000 miles and still be running. Of course, that doesn't change my opinion of it. It just makes it a well-maintained lousy car.

On the other hand, I may write a review saying that a particular car is a real gem and readers would do well to check it out when looking for a new car.

Next day the first call will be from someone who owns an older version of that same car.

"Listen," the caller will say, "I have that same car. It's five years old, it has 30,000 miles on it and it sounds like a gravel mixer and has less pep than a diesel."

My first question? "How often do you change the oil and filter?"

The answer that invariably comes back is: "7500 miles — sometimes — and the filter once a year, if it needs it. What's that to you? Detroit (or Japan or Germany or whatever the home country is of the neglected car in question) builds lousy cars and you know it. I'll never buy domestic (or foreign) again."

To have some fun I then ask: "And how often do you change the transmission filter?"

"Whaaa . . . What filter you talking about?" the caller replies.

What I have done is establish the fact that the caller spends little or no time maintaining his or her car but expects Detroit or Japan or whomever to have built a car that lasts 10 years or 100,000 miles without $5 worth of service or repairs. They don't. Nobody does. Or can.

I once had a young lady call with a problem on a Ford. I asked when was the last time she changed the oil. She said she hadn't had to change it yet. The book said owners should change oil every 10,000 miles and she drives less than 5000 miles a year and therefore hadn't reached 10,000 miles yet because the car wasn't quite two years old.

A man once responded to the oil change question by saying he changes oil when the oil light in the dash goes on. "Isn't that what that damn light is for?" he asked. No, it's not.

One of the best ones I've heard was from a Chevy dealer who claims to have removed an oil filter from a three year old car that actually weighed three pounds from the accumulated sludge that had hardened inside. He had to be exaggerating. I hope.

In each case, however, the car owner blames the car — not him- or herself — for the problem.

In most cases the calls come from people who own cars but know little if anything about their mechanical operation . . . that's most people. If it's you, then this is the book for you because I'm going to teach you a few basics about how to care for your car that will keep it running and help you to avoid some major expenses.

So you don't know what a piston looks like, let alone where to find one. I can't sew but I can thread a needle. And you can learn to perform enough tasks on your car to keep it running properly or to at least identify a problem so that the mechanic who knows how to do the work can make the repair in time to stop it from becoming a major problem.

An item that can be helpful to you in caring for your car is the owner's manual. This will lay out for you in both words and pictures all the parts and systems of your vehicle — what they are, how they work, where they are located. It will identify what can go wrong and what the symptoms are. And it will identify the particular replacement parts that are recommended for use in the vehicle.

There is an owner's manual issued with every new vehicle. When buying used, ask for it either at the time of inspection or at title-taking. If for some reason there is none, call or write to the zone office of the manufacturer. It will cost you some money but it is well worth the investment.

The time to start taking care of your car is the moment you own it. As soon as the sale is complete, head home and run it into the garage (or back alley if you don't have one). Change the oil and filter and any other fluids and filters so that you are starting off fresh. Record the dates and mileage the service was performed. Most stores with auto parts departments sell a package of door stickers like the ones car dealerships use to record service work. Put a sticker in your car door as a reminder of what work was done and when.

As for long-term care, there are three particular areas that require attention. They are oil and other fluids, tires and plugs. You need no more skill other than the ability to read, see and touch to take care of each.

CAR CARE
PART 2

Oil and other vital fluids

Oil is the lifeblood of your car. I know I said that before but I'm saying it again because I want to get the point across. Ignore it and your car's engine will have a life expectancy that no insurance company would touch.

There are some other vital fluids in cars, too—transmission, brake, power steering, coolant—and we'll talk about each of them in this chapter. But let's begin with oil.

Oil cools and lubricates your engine. Additives in the oil fight rust and corrosives. As you drive, the oil picks up pieces of rust and corroded metal and contaminants from the combustion process. As a result, the oil needs to be periodically changed along with the oil filter that traps most of these contaminants.

The rate of oil contamination and additive depletion varies with the type of driving that is being done. Unfortunately, the type that most of us do—the daily short haul, stop and start kind—doesn't warm the engine up properly and so increases the rate of contamination.

So how often should you change the oil? My own view is 1500 miles or every two months for both oil and filter . . . and I consider that to be a very cheap insurance policy to say the least.

Unfortunately, not everyone does. Too many car owners put it off for thousands of miles or months more than is wise.

And when they do get around to doing it, they try to do it on the cheap.

A few years ago a colleague at work bragged about finding a brand name oil at a discount store for something like three cans for a dollar. He naturally bought a case. I said the deal sounded too good to be true. I was right.

He brought a can to work and at first glance everything looked fine. The can, indeed, was a brand name, clean on the outside and not severely dented which I thought might have prompted the radically low price. Then I looked at the top of the can and realized why he had gotten such a good deal.

He had purchased a can of mineral oil. If he had put it into the car, the next stop would have been the repair shop — assuming he could have gotten it there.

Many people do not seem to realize that there are codes on oil cans that tell you whether you are buying the right oil for your car. There are two codes: one is called a service classification code and the other denotes viscosity — how thick the oil is and thus the speed at which it flows.

The service classification code used to be complex but oil, like tires, has undergone vast improvement over the past 25 years so now all you need to do is look for a can that says "SF" on the lid. SF oils contain all the additives and has all the anti-wear, high and low temperature stability, rust and anti-corrosion properties that any car's engine needs. And, best of all, it can be used in any vehicle regardless of make or model. So the simple rule is: No SF, No Buy. Follow it and you can't go wrong.

The viscosity code is slightly more complex. Viscosity is a measure of an oil's resistance to flow. Thick, slower flowing oils have higher numbers (10W-40) and thin, freer flowing oils have lower numbers (10W-20). Multi-grade oils (5W-40) have free flowing properties light enough for easy cranking at low temperatures and heavy enough to perform at high temperatures. The W means it's suitable for low temperatures or winter use.

For years 10W-30 was recommended for just about everything. Then in the early '80s, GM among others started advising motorists to use 10W-40 in its cars because it was supposed to offer more protection at higher temperatures.

In 1984 GM backed off. The reason was fuel economy. As a GM engineer explained to me 10W-40 is formulated with a thick, syrup-like viscosity improver that added more deposits in newer engines at higher temperatures. The fear was oil deposits in the rings and sticking rings that could cause engine damage. Because 10W-30's viscosity was thinner, there would be less drag or friction on internal parts which would translate into better fuel economy.

As a rule, GM now recommends 10W-30 oil in V-8 engines and 5W-30 in fours and sixes.

You will find the recommended oil for your car in the owners manual. It's best to look it up and follow it.

Detroit, led by General Motors, finally realized that they are helping both car owners and themselves by promoting more frequent oil and filter changes. It only took them several decades to see the light.

Do you economize by eating breakfast, lunch and dinner off the same plate? Well, for years Detroit tried to promote a similar concept with its ruse of "extended service intervals."

In the early '80s, as I pointed out earlier, Detroit was in the process of downsizing cars, developing new engines, boosting fuel economy—all projects that required multibillion dollar investments. The result was that the prices on the window stickers on new cars rose like a helium balloon. Consumers in turn reacted by not buying if they didn't have to. The industry sold more than eight million cars annually between 1976 and 1979. But as prices went up, sales went down and the number of cars sold dropped below seven million between 1980 and 1983.

That's when Detroit came up with the idea of extended service intervals and told consumers that they could go a minimum of 7500 miles and often 10,000 to 12,000 miles between oil changes.

"We may have raised prices by $1000 a year but look at all the dough you're saving by only changing oil once a year," Detroit seemed to be saying. Though the difference between a $10 oil change and a $1000 boost in the window sticker hardly offset each other, Detroit had a field day promoting extended service intervals.

Detroit gained in another way. Tell people they don't have to change oil for 10,000 miles and they'll try to get 15,000 miles out of that oil. Well, with new car warranties in 1980–1983 at 12 months or 12,000 miles, by the time motorists started developing engine trouble, the warranty period was over and they were stuck making their own repairs. The consumer saved nothing.

In April, 1984, a major change occurred—quietly. GM brought out its downsized C-body cars—the Olds 98, Buick Electra, and Cadillac DeVille—and called them 1985 models. With no fanfare they stuck a little booklet in the glove box of each car. I test drove one of the cars and found the booklet and couldn't believe what I saw. The booklet was entitled, "Are you changing your oil often enough?"

That booklet represented a greater departure in the way GM was doing things than the fact it had taken a foot in length and 800 pounds in weight out of the 98, Electra and DeVille and converted the trio to front-wheel drive.

That was the first reaction by a member of the industry to criticism that had been building over extended service intervals as consumers had begun to realize that the only thing being extended was their repair bills. Extended service intervals, such as delayed oil changes, were giving motorists a false sense of security, resulting in less vehicle maintenance and higher and more frequent repair bills.

For years GM, like the others, promoted oil changes at 7500 or more miles and filter changes at every other oil change. A little asterisk accompanied the oil change information for those who bothered to skip to the fine print at the bottom of the page that stated a 7500 mile oil change was advised for "normal" driving, but a more frequent oil change was recommended under "severe" driving conditions.

The problem was that normal driving was defined as lots of daily highway driving where the engine and all the fluids got warmed up and no condensation formed, whereas severe driving was defined as the stop and start, short haul driving that never gave the engine or fluids a chance to warm up.

"Normal" driving actually was abnormal, the type performed by salesmen who put lots of "good" miles on the car each day: long, straight hauls where all fluids got up to proper operating temperature.

In truth, "severe" was actually the "normal" driving pattern performed by most motorists each day. The "severe" driver would pull the dipstick and the reading would be at or near the "full" mark and he or she would think everything was fine. In fact, however, the reason the reading was "full" was the condensation in the crankcase that helped push the oil up to the mark.

The Automotive Filter Manufacturers Council once estimated that 20 percent of drivers operated their cars under the so-called "normal" operating pattern, whereas 80 percent or four out of every five of us drove our cars under the "severe" pattern.

GM, then, informed its owners in the booklet to change the oil and filter at 3000 miles or three months, whichever comes first, if you make frequent short trips of less than four miles, experience frequent stopping and starting such as in rush hour traffic, or operate your car mostly at low speeds. If you run your car daily for several miles at a time and none of the above conditions apply, then GM says a 7500 mile change is in order.

What about the so-called "super oils," you ask—those that say you can leave them in your engine for 10,000 or 20,000 or more miles without the need for a change? The bargain hunter sees 20,000 miles of usage and is led to believe he or she is actually saving money on a $3 to $5 per quart container of oil.

The lazy ones among us leap at the chance to run a car for a year without an oil change and will use the oil. One of the problems is that if the oil says don't change it for 10,000 miles, the motorist will believe that to be a minimum, not a maximum, and wait for 15,000 miles to get even more use out of it since it cost so much to begin with.

I've found a 79-cent can of Quaker State oil to be super—providing I add a fresh supply every 1500 to 2000 miles and change the filter. Whatever the brand, if the can says SF and you change with regularity, you'll prolong the life and guarantee the proper operation of your car for thousands of miles.

In addition to the oil dipstick, there's another dipstick under the hood which gives you a reading on the transmission fluid level. The transmission fluid operates a series of pumps and valves that enables you to shift gears. The fluid also lubricates and cools the transmission.

The owner's manual will provide a diagram on how to read the dipstick to ensure that the fluid level is up to par. To be certain of getting an accurate reading, always check the level after the car has been warmed up. In refilling, use only the type of fluid recommended in the manual—don't economize with a bargain basement brand. And *never* overfill.

Not only is it important that you have ample fluid, it's vital you examine the color of the liquid. Transmission fluid should be pink, reddish or maroon in color. If it's orange or brownish and smells burnt, you've got trouble that a simple change alone might not cure. Off to a mechanic quickly.

There's also a transmission filter which traps particles and waste contaminants. The reason to change the fluid and filter is to rid the system of those impurities. Your owner's manual will provide you with the recommended transmission fluid and filter change intervals. Most mechanics advise doing so at about 24,000 miles. Some carmakers say you can get by with 100,000 miles. I favor the lower mileage interval.

Coolant protects the engine year round despite the fact it commonly is referred to as antifreeze which suggests only wintertime importance.

The function of the cooling system is to keep your engine at the proper operating temperature and to do so requires a mix of water and ethylene glycol which most call antifreeze but should call coolant. Coolant or antifreeze, its job is to keep the engine from freezing in the winter, boiling in the summer. Your owner's manual and the information on the coolant container will tell you the proper water/coolant mix to add to your cooling system. Never add just water or just coolant.

Thanks to today's use of plastic reservoirs or "burp" tanks, you can tell if the coolant level is proper at a glance. Those tanks have lines for "cold" and "hot" level checks. It used to be you had to remove the radiator cap, not always an easy or pleasant chore after the car's been run awhile. Now you need only pop the top of the plastic container and add coolant.

However, making the task of checking coolant level so easy has led to some problems. No one opens the radiator cap anymore for a visual inspection of the fluid or the radiator core. Just because the coolant still looks greenish yellow after two years doesn't mean it's still working. Coolant contains additives to combat corrosives in the system and those additives break down and dissolve from heat. If the coolant isn't flushed every two years and new, fresh fluid put in, you're asking for trouble. If you pop the radiator cap and see sludge buildup on the core or solid matter floating around, flush the system at once.

Warning: If you don't have a reservoir tank or want to open the radiator cap to inspect for sludge, only do so when the engine is cool.

Brake and power steering fluid each are housed in their

own little containers. While the containers are small, the job of each fluid is vital. Each should be checked periodically, if not monthly at least every six months and certainly at the annual tune-up.

Brake fluid, in very simple terms, activates the brakes to stop the wheels. The brake master cylinder contains a hydraulic fluid which under pressure causes each wheel cylinder to stop the wheel it's connected to. The fluid evaporates gradually and should be checked periodically. If you find yourself having to pump the brakes to stop or if you suddenly have to push the brake pedal lower to the floor than normal to stop, have the fluid and the brake system checked at once. The brake fluid could be low or a brake adjustment needed. The location of the brake fluid container is pointed out in your owner's manual. Many newer cars have plastic see-through reservoirs for a quick visual inspection. Be careful of the fluid, it's corrosive and if you spill a drop on the car body, the finish will carry the mark forever. If you find the fluid very low and have to add a considerable amount, you probably have a leak. Have your mechanic check out the entire system for possible troubles. If you get air in the system, the air will have to be "bled" and the shadetree mechanic should leave that to the pros.

The power steering system uses hydraulic fluid under pressure to help you turn those wheels. Again the manual will tell you where the reservoir container is that holds the little power steering fluid level dipstick. If you haven't been checking the fluid level, detecting sluggish steering will remind you of the neglect. If the fluid level is low, add to the full mark. As with brake fluid reservoirs, many new cars have see-through power steering system containers for handy visual check.

CAR CARE
PART 3

Tires

All season tires, those treads that can be kept on your car year 'round without needing to be changed with the seasons, have led to one of those good news, bad news scenarios.

The good news is that if you live in the snow-belt you don't have to put up with freezing your hands as you fumble to take off the summer tires and put on snow treads when there's six inches of snow on the ground.

The bad news is that since you don't have to worry about changing them with the seasons, you don't spend the few minutes late each fall and early each spring inspecting the tires during the biannual changeover.

Tires, like spark plugs, give us clues that wear patterns are developing that should be remedied before we have to invest a week's pay in a new set of treads. They also give us warning that we have mechanical troubles brewing with the car. Those mechanical problems, if left uncorrected, not only will cost us a set of tires, they will result in major and costly work to be done — or, worse yet, a breakdown on the road-way when we least need or expect it.

In recent years the old bias ply tires have been replaced by radial ply tires. Radials, if treated properly, will give 40,000 to 60,000 or more miles of driving. That's real value for the money. But without proper care you'll get 10,000 to

Trouble Signs

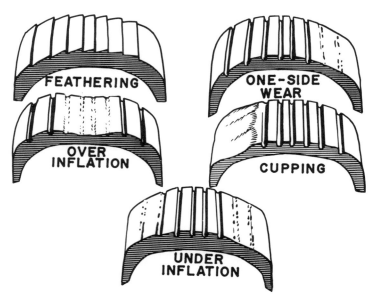

20,000 miles, if that much. If you're paying for radial tire technology, get the benefit of it and don't settle for the same mileage you used to get out of the bias ply tires. Radials will wear out just as quickly as the old tires if you neglect them.

Signs of tire problems are under- and over-inflation, feathering, one-sided wear and cupping. Actually, under- and over-inflation are both signs and causes of tire problems.

Under-inflation is probably the most common problem and is generally caused by one of several factors. Many drivers simply forget to check for tire inflation. Others, who intend to inflate their tires, find that the gas station now charges for air and refuse to pay for what they can breathe free. The other factor is that even when we do check them and fill them we tend to forget that tires now require much higher pressures than they did in the '60s and '70s.

When Detroit started its downsizing campaign to make

cars lighter and more fuel efficient in order to meet federal fuel economy legislation, it asked Akron, home of the tire industry, for some help.

The tire makers responded with steel belted radial ply tires as well as low profile so-called P-metric tires, ones that could hold 30 to 35 pounds of air. The more air, the better rolling resistance or ability to almost glide the car down the roadway.

When I started driving cars 24 pounds of air was normal, 28 pounds when you were hauling a big load. Many motorists remember what they *used* to put in tires and still inflate tires to 24 or 28 pounds when the sidewall specifies 30, 32 or 36 pounds for optimum performance. So in trying to care for the tires many of us are unknowingly underinflating our tires.

Because of federal regulations, there's now so much writing on a tire sidewall that you may spend 10 minutes trying to find those tiny numbers that give maximum inflation pressure. It's worth the hunt. Find out what pressure your manufacturer recommends and then inflate to that amount and keep it there.

And always check your inflation pressures when the tires are cool. A tire's air pressure automatically increases as its internal temperature increases and that happens when it's driven.

When a tire is under inflated most of the contact with the road is on the outer tread ribs, which causes them to wear faster than the middle of the tire. With *over-inflation*, which occurs when you put too much air in the tires, it is the middle of the tread which gets most of the action against the road and the inner ribs wear much faster than the outer ribs.

Neither under- nor over-inflation are signs of mechanical problems. But either one can and does have an effect on vehicle handling, cornering and braking — and the effect generally is an adverse one. Cars are designed for use with specific sized tires at certain inflation pressures. The tires and suspension systems work in tandem. By under- or overinflating your tires you are upsetting the operational balance the manufacturer had in mind.

Feathering is a condition in which the edges of the tread ribs take on the appearance of feathers. It's caused by erratic scrubbing against the road when a tire is in need of toe-in or toe-out alignment correction.

Another alignment problem, *excessive camber*, results in an outer rib or shoulder of the tire wearing down faster than the rest of the tire. It means the tire is leaning too much to the inside or outside of the tread, placing all the burden of the tire's work on one side or the other.

Cupping is one wear pattern you hope not to see. Dips or cups in the tread are a sign that the wheels are out of balance or the shocks are worn or the ball joints need replacing. Most cars built in the '80s have ball joint wear indicators. If a service shop puts your car up on the rack and wiggles the tire and says: "See, you need new ball joints," you reply: "But that's what the tire is supposed to do, let's look at the ball joint wear indicators." Actually your best bet is to have him lower the rack and get out.

Two other reminders: never mix and match different tire constructions on your car and follow the manufacturer's recommended rotation to ensure the longest life for your treads.

With regard to tire rotation, things have changed over the years. When steel belted radials first appeared in the '70s, the tiremakers advised any rotation be done carefully so that tires on the right side always stay on the right, the left on the left. Those early radials developed specific wear patterns based on what side of the car they first were on. They "set" in a particular position. As a result, the early rotation pattern on radials was simply front to back and back to front and keep them on the same side.

I checked with Goodyear Tire & Rubber Company, Firestone Tire and Rubber Company and Bridgestone Tire and Rubber Company, and here's what they all now recommend.

Tire Rotation Patterns

FRONT-WHEEL-DRIVE VEHICLES   **REAR-WHEEL-DRIVE VEHICLES**

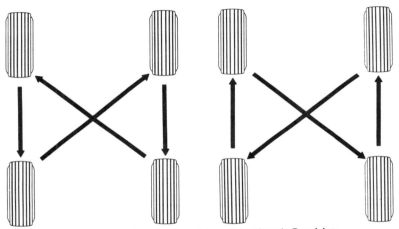

In both cases, "X" non-drive-axle tires to the drive axle. Remaining tires go straight to the front or rear accordingly.

If you are driving a rear-wheel drive car, rotate the rear tires to the front, left rear to left front, right rear to right front. Move the front tires to the rear, but criss-cross them, right front to left rear, left front to right rear.

On a front-wheel drive car, you first move the front tires to the rear wheels, left front to left rear, right front to right rear. Then you move the rear tires to the front and criss-cross them, left rear to right front, right rear to left front.

Check your tire manual on how often the rotation is advised, usually 5000 to 6000 mile intervals, but that will differ by manufacturer.

Wait, you say, you haven't mentioned retreads or over-sized tires. I didn't mention retreads for five good reasons: (1) my wife Sue, (2) my son Brad, (3) my daughter Gina, (4) my daughter Dana and (5) myself. I don't want any of them riding on retreads.

But aren't they so much cheaper than all-new tires? Right. Over a year's time you'll save three cents a day in the difference in price between new tires and retreads. That's $10.95 a year . . . $10.98 in leap year. It's not worth it. The risk is far greater than the savings.

As for oversized tires, if you mean those monsters you see on pickup trucks or 4x4's, what you gain off the road you lose on the road in really miserable ride and handling.

I've driven utility vehicles with oversized tires and never enter a sharp turn or corner without both hands firmly locked — in prayer. There is very poor cornering stability with oversized tires. And power steering compounds the problem. When you add the fact that many of the 4x4's sit high with a raised center of gravity, oversized tires serve to upset the balance even more. They scare me.

But if oversized to you means stepping up one size from the tires that came on your car, that's different. Providing you have the clearance with the suspension, moving up one size often will improve ride and handling — but usually in a small subcompact or compact that started out offering undersized tires to begin with.

When in doubt, check with your car manufacturer and your tire maker for their recommendations.

CAR CARE
PART 4

The car hasn't been acting right. You take it to the corner garage, local independent mechanic or a dealership to have the engine tuned. You pick the car up, note that the bill says you paid for a fresh set of four, six or eight new spark plugs, and off you go, content that the engine is perky and peppy once again.

The mechanic has done you wrong. But don't place all the blame on him. You're guilty, too, of not seeing the four, six or eight plugs that were removed and going over what troubles you still have despite having a fresh set of plugs in place.

A close examination of the old spark plugs can provide clues to more serious problems you might have. But since a plug is small, relatively inexpensive (far less than a rebuilt carburetor at least), and so easy to replace, most motorists pay little or no attention to them.

Admittedly, I've never walked up to people at a party and said: "How're your spark plugs, they tell you anything?" But those old plugs should be given a close examination if you do your own work, or by your mechanic if you farm the job out. If someone else does the work, he or she should tell you what if any problems were found when taking the old plugs out and putting the new ones in.

A spark plug has no moving parts. Its function is to

insulate the passage of thousands of volts of electricity into the combustion chamber to ignite the air-fuel mix. The firing tip is exposed to temperatures in excess of 3000 degrees.

Because a plug is associated with ignition, it plays a major role in fuel economy and emissions. If all the fuel isn't being burned, you're not getting optimum fuel economy and hydrocarbons are being formed.

So it's a small but vital part of your engine. And it's a tattletale, too. Your car's plugs can tell a mechanic that piston rings are worn; the fuel/air mix is too lean or too rich; timing is overly advanced; valves are sticking; coolant level is low; intake manifold is leaking and a variety of other ills.

To check for trouble signs, you have to remove the plugs. When you do, mark each so that you know which cylinder the plug was taken from. Knowing the position can help detect a problem. One example would be the two front plugs being fouled on a six cylinder engine. That could indicate a blown headgasket and not just excessive mileage between tunes.

Being able to read warning signs has become more important as engines have gotten smaller. When one plug on a V-8 wasn't firing, chances are you didn't notice a loss of one-eighth of your performance. When one plug on a four cylinder isn't working, you're losing 25 percent of your power and you'll notice.

A *normal plug* (see spark plug illustrations at end of chapter) has brown or grayish tan deposits on the insulator tip. You'll detect slight electrode wear which means the plug has the correct heat range for your engine. Clean, regap and replace the plug.

An *oil fouled plug* will have black oil deposits on the insulator tip and electrodes and means an excessive amount of oil is entering the combustion chamber. This could mean defective piston rings or valve seals which cause the engine to consume too much oil. If the engine has a lot of miles it could mean the piston rings or cylinder walls may be worn excessively. It may mean time for a new $1200 engine.

If you have an overhead valve engine, the oil may be seeping past the valve guides while if the fuel pump diaphragm is ruptured, oil may be entering the combustion chamber to foul the plugs. Some may treat the problem by installing a hot spark plug to burn off the oil. That's treating the symptom rather than finding a cure. A healthy spark plug doesn't make a sick engine healthy.

A *carbon fouled spark plug* will have dry, fluffy black deposits. That could mean you have the wrong plug for your engine, a plug with too low a heat range. However, it also could mean that you're burning too rich an air/fuel mix, either because the choke is sticking, the air cleaner clogged, timing retarded, compression too low, or you make several short, slow, stop and go trips a day that help build up deposits.

If you have a car that still uses leaded gas, or have doctored the fuel tank inlet to accept leaded gas, you could have *lead fouled plugs*, detectable by white, brown or yellow deposits on the electrodes. If you are purchasing a used car and aren't sure if the previous owner used leaded gas when he shouldn't (and ruined the catalytic converter which you must replace at your expense to pass your state's emissions test), checking for a lead fouled plug certainly will provide you with some evidence.

A *worn out plug* will have a rounded electrode that's well worn. The voltage to spark the gap has increased — perhaps even doubled from normal. You're wasting power and fuel and increasing emissions. If you are looking at a used car that has 10,000 miles on it and the plug looks like its on its third 10,000 miles, you have another clue you may not be buying what you see.

Detonation and pre-ignition can cause severe damage to your engine. Detonation can break the insulator core nose of the plug. Detonation occurs when some of the air/fuel mix begins to burn on its own spontaneously from increased heat and pressure just after ignition. The resulting explosion puts pressure on internal engine parts and the increased heat can cause pre-ignition.

Detonation can occur when ignition timing is too far advanced, the octane of the gas you use is too low, the air/fuel mix is too lean, carburetion is poor or there are leaks in the intake manifold, or deposits in the combustion chamber have raised the compression ratio.

With *pre-ignition* the electrode is burned and the insulator blistered. Pre-ignition occurs from ignition of the fuel charge before the timed spark. Hot spots in the combustion chamber could initiate combustion. Among the causes are combustion chamber deposits, those hot spots in the combustion chamber from poor control of engine heat, piston scuffing from inadequate lubrication, detonation mentioned above, cross firing from an electrical induction between spark plug wires, or spark plug heat range too high for the engine.

Simply changing the plugs will do little good without pinpointing and resolving the problem. And if you spot either of these two problems when checking out a used car, you probably will want to keep looking—for another car.

An *overheated plug* will have a clean white insulator core nose and/or excessive electrode erosion. You may also find the insulator blistered. That could mean no more than a plug that is too hot for the engine. But it also could be a sign that ignition timing is overadvanced, there is too little coolant in the cooling system or a blockage in the cooling system such as a radiator filled to the brim with sludge, too lean an air/fuel mix, or a leaking intake manifold that is causing the plug to overheat.

Ash fouling occurs when oil and/or gas additives are burning during the normal combustion process. They will appear as light brown or white crusty deposits on the side or center electrodes and core nose. If the deposits are excessive they'll mask the spark and the engine will misfire.

Glazing is a shiny deposit that is yellow or tan in color. Its presence indicates that plug temperatures have suddenly risen during a hard, fast acceleration and deposits melted to form a conductive coating which can cause a misfire. Glazing can't be removed. You have to replace the plugs. It usually means plugs should be of a colder temperature range.

Finally, you may spot a bent electrode and a broken insulator, a sign of *mechanical damage* resulting from a foreign object falling into the combustion chamber.

In a six cylinder engine, if the two front plugs have become fouled, it may mean a blown head gasket and not simply excessive mileage between tunes.

If the two center plugs have become fouled, raw gas may have spewed out of the carburetor and into the intake manifold after the engine was shut off. If you do lots of stop-start, short haul driving, the center plugs may have become fouled from an overly rich fuel/air mix.

In an eight cylinder engine, if the four back plugs show signs of overheating (blistered white insulator, eroded electrodes), you may have cooling system problems. Coolant probably isn't circulating to the rear of the engine, which would cure the problem.

If only one plug is overheated, it may mean an intake manifold leak near the cylinder. On an eight you might find the two rear plugs oil fouled, especially if the car has been subjected to start-stop, short haul driving. The oil drain holes in the rear of the cylinder head may have become plugged from sludge, and oil may have been pulled in around the intake valve stems. High oil consumption and a smoky exhaust are other clues.

A new plug requires about 5000 to 8000 volts to fire but after 10,000 miles it could take double that.

If you have just purchased or someday will buy a new car and have trouble with starts and stalls you could be a victim of cold fouling which results when the car has been started and stopped several times between assembly and sale on the showroom floor. These quick starts include getting the car off the assembly line, to the parking lot, onto the carrier, off the truck at the showroom, to the wash rack and back, onto the floor, and out for a test drive.

The fix is easy: replace those new plugs with a fresh set.

Lead Fouled

Normal Plug

Oil Fouled

Carbon Fouled

Worn Out

Detonation

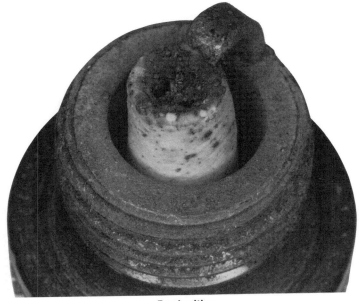

Pre-ignition

Overheated

Ash Fouling

Glazing

Mechanical Damage

CAR CARE
PART 5

Bogus parts

I vividly remember an episode that took place during my eighth grade year. It was recess and all of us were on the playground. A friend was roaming about showing off a $10 bill. Each classmate that saw the bill broke into laughter.

I finally gave in and asked to see the bill. He agreed and I looked at it for several seconds. Nothing struck me as funny, until he realized the word hadn't gotten around to me yet.

"Look at the signature, dummy," he said.

I did. And right there under Secretary of the Treasury was written Michael "Mickey" Mouse. No wonder they call it funny money.

Maybe Mickey Mouse sawbucks are worth a chuckle, but when it comes to counterfeit or bogus automotive replacement parts and components there is no humor. Pick up a bogus $10 bill at a store and you're out $10 (unless you pass it on to some other sucker very quickly). Pick up a bogus oil filter and you could be out $2000 for an engine. Or, worse yet, pick up a set of phony brakes and you could lose your life.

Unknowingly you may have had at one time, and perhaps even now, a bogus part or component in your car. The estimates of just how big the market really is for counterfeit parts varies by who does the estimating, but I've been told annual sales in the U.S. alone run from $3 to $6 billion. Repeat. That's billion with a "b" as in *beware.*

Whether bogus, counterfeit or simply look-alike parts, the scheme is the same. A cut-rate parts manufacturer tries to make you think you're getting the brand name part and the quality associated with it.

Here's how the deception works. You walk into a parts store, gas station or even a mass merchandiser to pick up an oil filter for your Chevy Caprice or Ford LTD or Chrysler New Yorker. You are familiar with the AC-Delco package for the GM oil filter or the Motorcraft box for the Ford filter or the Mopar package for the Chrysler replacement parts. You drive an LTD so you ask the counterman for a Ford filter and he brings one out, you pay him for it, and off you go for home and the chore of replacing the used filter with a fresh one to keep that LTD running properly.

At home you remove the old filter, put it in the box the new filter came in with the familiar Motorcraft logo with the picture of the speeding car on it. You dispose of the box with the old filter in it. The new filter is installed and the job is done. A week or two later the car just isn't running right. You look on the driveway and there's a puddle of oil. You start the car to move it into the garage to survey the damage and "ping," the engine just threw a rod.

The car only has 10,000 miles on the odometer. You call your dealer and arrange a tow to his shop. While he's working on your car you contact Ford to raise a stink. How could you throw a rod after only 10,000 miles of driving, especially when you've religiously changed oil every 1500 miles?

Not only do you make a fuss with the dealer, you write or call Dearborn, Ford's corporate headquarters. In fact, riled beyond belief, you contact the National Highway Traffic Safety Administration and even call the Center for Auto Safety in Washington, the Ralph Nader-inspired consumer watchdog group.

At work you tell your colleagues about that "lemon"

Ford you own. At home you let all your neighbors in on how lousy Ford engines are. You write your friendly auto editor for the local newspaper in order to try to get him to do a feature story on Ford engines that throw rods after only 10,000 miles.

Ford sends out a representative to inspect the car and its engine — or what's left of it. He finds the problem. It was the oil filter, or at least what looked like a filter. When the cannister is cut open and the excess oil wiped away you can see a series of words written in an Oriental language. Alongside the words is a picture — of bean sprouts. The oil filter was made out of a discarded vegetable can from Korea, Taiwan, Hong Kong or the like.

You head for the garbage. Fortunately the box the filter came in is still there, though dripping with oil. (This is a rarity. Usually, the garbage has been picked up and the evidence is long gone. In fact, the outfit that made the product is counting on the fact you've discarded the box weeks ago before any trouble pops up and you start looking for clues). Now, for the first time, you realize what happened. The logo on the box doesn't say Motorcraft to distinguish it as a Ford made product. It says Motorcare (or Motorcar, another name used on the look-alike packages).

The vegetable tin couldn't withstand the heat or pressure of pretending to be a filter. The flimsy paper element inside disintegrated and dislodged in the engine. The damage: one engine — $2000. Is Ford liable? Certainly not. You can take the box and filter and receipt back to where you got it and demand restitution. But can you prove the filter came from there? You have a box covered with oil, but the receipt is long gone. And in most cases the victim kept neither the box nor the receipt.

In fact, in many cases the motorist who experiences engine damage doesn't suspect the filter. He might not have problems for three months after installing one of the bogus parts and will have forgotten all about the filter. Perhaps the car isn't worth the investment in an engine and he junks the car — evidence and all.

Parts like this oil filter typically are made in Taiwan, Hong Kong or Korea. Once here, these cheapies are then put in boxes made to resemble the quality product in which the manufacturer has invested millions to develop, market and warrant. Deception at its worst. Preying on the good name of a proven producer of quality parts.

A Ford Motorcraft look-alike, as I said, may read Motorcare or Motorcar. I've been told of boxes that in bold letters read "FOOD," a planned misspelling of "FORD." The AC-Delco name sometimes comes out just AC in the familiar circle logo or just Delco or AC-Deltco, another misspelling.

Names, words, logos and colorings are so similar to the original you wouldn't suspect anything until it's too late and the part is in your car and the packages in the garbage. It can be a tragic optical illusion.

No part is immune to those who prey on the good name of the genuine parts manufacturer. There have been reports of replacement gas caps with no federally required shut off valve to prevent fuel leakage in a rollover accident. Using one of those caps would be like putting a fuse in your tank — and lighting it.

Such simple and inexpensive parts as turn signals and warning flashers have been known to be counterfeited as well. Imagine thinking you've signaled a left hand turn and proceed into a line of traffic or thinking those flashers that aren't working are warning cars approaching from behind that you're stalled in the roadway.

It's a difficult practice to stop because it is such a lucrative trade. If a parts outlet or filling station normally pays $3 for a part but is offered the same part for $1 each, well, you can see the motivation. And to show you the greed in the world, oftentimes the parts store that saved $2 on the part will turn around and sell it for the same retail price he normally sells the $3 wholesale part for.

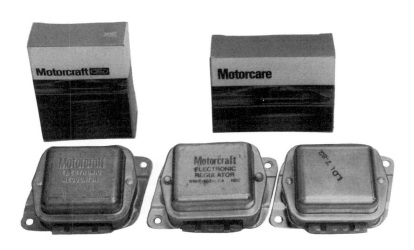

BOGUS BOXES FOR BOGUS AUTOMOTIVE PARTS—Counterfeit automotive parts cheat the manufacturer, the dealer and the customer. The box from a genuine Motorcraft electronic voltage regulator (left) is shown beside a phony—a simulated "Motorcare" box designed to appear like the original—which contained an unlabeled regulator. Motorcraft is a trademark of Ford Motor Company. Both Ford and Motorcraft automotive parts are sold by Ford Parts and Service Division.

There's another problem in trying to curb this illicit trade. The parts manufacturers sought and got government laws enacted to curb parts trafficking. In 1984 the government enacted the Anti-Counterfeit Act which made manufacturing and selling counterfeit parts a felony punishable by up to five years in prison and fines of up to $1 million. But there was a loophole.

If an outfit in Hong Kong puts a vegetable can oil filter in a box that reads "Motorcraft," it's a counterfeit under the law and subject to criminal prosecution. But if the manufacturer puts a vegetable can oil filter in a box that reads "Motorcare" it then becomes *only* a simulation or look-alike packaging and subject only to a civil penalty and could take years to get on the court docket before the culprit is slapped on the hands.

During the three years it might take for the case to be heard the outfit is still selling its "Motorcare" filters. Finally the court hears the case and orders the outfit to stop distributing "Motorcare" filters. It does, but the next day starts selling its filters in "Motorcar" boxes and the legal process starts all over again. Guess who pays for the genuine manufacturer to fight the simulator? You and me in the price we pay for our genuine parts.

What can you or I do as consumers to stop this practice and avoid potential monetary or, worse yet, physical damage to ourselves and our families? We can start inspecting the parts packages we buy more closely. We can ask for the replacement parts by name and familiarize ourselves with the packaging. When the counterman brings out a box that says Motorcare we can say, "Wait a minute, that's not what I asked for, that's not a genuine part. Why are you carrying it?"

Ford, GM, Chrysler—all major manufacturers will be eager to hear from you when you spot such a package.

What if you take your car to a shop to be repaired and leave it for the day? You demand to see what parts will be used and what packages they come in. If the mechanic pulls out a Motorcare box you get in your car and pull out—and report the episode to Ford.

The business will prosper as long as people continue buying these parts. Those who sell the bogus parts will continue to do so as long as they go unchecked. The monetary considerations are great but the store that sells this junk won't make much of a profit if its customers refuse to buy there.

The trade in parts has been so lucrative, it now has spread to body components — fenders, doors or quarter panels. Again, the body parts come from a foreign source, the bulk from Taiwan. Instead of an oil filter, say you damaged a fender. You pick one up at a parts outlet and put it on. In a few months the fender starts to rust. Some would complain. Others would write it off as an after-effect of the accident.

What it could be is an "imitation" body panel that only looks like the fender replacement for your vehicle. Trouble is there's been no protective coatings applied to fight off corrosion like the factory manufacturer would do. And while the genuine fender manufacturer uses an expensive die to produce his fenders and only stamps out 15,000 fenders off that one die, the producer of the imitation fender uses a soft metal die and stamps out 30,000 units. The tolerances just aren't there.

"Many of these items are deficient in fit and finish and very poor in rust resistance," says Kenneth Myers, product and marketing manager for Ford Motor Co.'s Parts and Service Division.

Worse yet, many in the insurance industry are helping to foster the spread of low cost (often 30 percent less than the genuine original) body panels.

When a car is damaged the insurance company wants to replace parts as cheaply as it can and Myers, for one, says they have taken to not just okaying, but actually recommending "economy sheet metal" replacements for damaged body parts.

If a Ford fender would cost $100 and the Taiwan-made fender $70, the $70 fender is put on. When it rusts or doesn't line up properly, the car owner blames Ford.

Wait, you question what an "economy" fender is and the shop tells you it comes from Taiwan. You call your insurance agent to roust him from the golf course.

"Sure, go ahead and use the genuine Ford part," he tells you after listening to you gripe for ten minutes. "But," he adds, "you'll have to pay the difference between the $70 and the $100." In other words, $30 comes out of your pocket.

How do you stop the practice? It won't be easy because as long as there's a $70 fender out there for the insurance adjusters to use as a benchmark, they don't care if you go with the cheapie or the real thing because anything over $70 doesn't come out of their till.

Ford says "the best" of the "economy" fenders they've tested withstood only 100 hours of a salt bath before rusting compared with a 500 hour salt bath its own fenders must withstand. So how do you avoid the grief and expense of getting one of the cheap replacements?

You specify to the person or shop making the replacement to use the genuine part and while it unfortunately will cost you extra money, when the agent gives you a choice, you go with quality. If possible, the next time you pay your insurance premium, fill the envelope with dollars — *Taiwan* dollars.

Some Tips

Asterisks to erotica

Be cautious of what I call the asterisk ads in which you may see fantastic claims followed by the familiar *. Read carefully the explanation as to what the * means. It could mean look out, watch out or get out.

If you see an ad for a sharp used car at a very low price and an * follows, it may state that the reason the '84 Mustang is priced at $3500 is that it's $3500 after you've put down a $3000 deposit. Or it may say "with approved credit." God could get the credit—maybe—but not you.

The * might also point out that the '84 Mustang is selling for $3500 "with suitable trade-in," meaning if you are willing to part with your '86 Town Car.

The worst abuse of the * in the market today is by dealers advertising a car for "$300 a month*" and then when you find the explanation of * it says: "With 10 percent down and 48 monthly *lease* payments." You think you're buying a car and instead you are leasing it. The dealer draws you in, and you don't want to lease, but with the same fast arithmetic he *sells* you a car.

Bait and switch is most common on new cars, such as the ad that boasts of the $9800 Cadillac. You travel to the dealership and there is no $9800 Cadillac but he can put you in a $15,000 one, a used one or a demo. You could fall victim to the bait and switch in used cars, too, when you go looking for that $3500 '84 Mustang and it was "just sold" but "I have another just like it for only . . ."

Insurance Insights

You bought a new car five years ago, had it insured, and every six to 12 months get a statement from the insurance company and pay the premium like clockwork. Next time the bill comes take a closer look at your coverage. The car was new so you got full coverage and probably either no collision deductible or a very small $50 deductible. You pay, perhaps, $300 every six months. But the car is now five years old. Go to a $250 or $500 deductible (after all, what's the car now worth?) and you could find that by accepting more of the risk, your premiums may drop to $150 every six months.

Be prepared. The agent probably will say, "Great, you're smart to raise your deductible because now you can apply the savings toward raising your liability coverage from $100,000 to $300,000." He's obviously looking to keep your money in his family, but he may be right. In today's litigious society, you may want to raise liability coverage to protect yourself.

Dear Diary

Want an early clue a problem may be developing under the hood? Keep a mileage diary in the glove box. Every time you fill up, record the mileage you got with the last tank of gas. The records may show you've been getting 20 miles per gallon consistently, but on the last tank you dropped to 18. It could have been you did a lot more stop and start driving than usual. But if you go from a consistent 20 to 18, then 16 and then 14, something is amiss. Fuel filter, air filter, perhaps under inflated tires, or too long since the last tune-up.

Years ago Ford Motor Company provided me with a car equipped with a computer that gave mileage averages for any length of trip. I took to a stretch of road, filled with gas and did two ten-mile road trip runs. I used the computer to get two average mileage readings that only varied by one mile per gallon from each other.

I then made the run with a towel lodged into the air cleaner nozzle to prevent a free flow of air to the engine and therefore a too rich fuel to air mix. The mileage dropped by six miles per gallon. Another run I let eight pounds of air out of the tires. Mileage dropped by four miles per gallon. On one run I even kept all the windows wide open just to see what so-called wind resistance would mean — a two miles per gallon loss in mileage.

By keeping a mileage diary you can spot minor problems before they become major (and costly) ones.

Clean Hands

Suppose you like to work on your own car, but get tired of the chore of cleaning your hands. You've used the bulky nylon or cotton gloves but they just get in the way. Get yourself some surgical gloves. They fit snug so you don't loose the feel of the wrench or screwdriver or filter. At the same time, you don't get so filthy you have to use a container of soap to get clean.

Instant Rustproofing

Your wagon has a luggage rack or your coupe one of those deck lid carriers. Hard to get in and clean around those things, the roof luggage rack especially. Most manufacturers put all the galvanized and zincrometals in the lower door panels, rocker panels and wheel wells so the roof doesn't get much rust protection. As a result, rust forms in the holes drilled in the roof to attach the metal screws of the rack. What you can do is undo the screws a few turns, shoot a light spray of oil on the screw, and refasten. Repeat it periodically to ensure continued protection.

Rustproofing II

Your car looks great except for that one bad fender. You finally get up the nerve to take the old one off and put a replacement on. Before the replacement is put on, take an oil can and saturate the inner metal. Leave the panel where the excess can drip off before you put it on. I once owned a Vega and replacing fenders was a yearly chore just like changing the snow treads — until I applied the oil. Not one rust mark or spot after two years, when I finally sold the car because the hood had rusted off the hinges.

Smoke Control

Got a smoker in the car, perhaps yourself? The car stinks from the smoke after a while. And the windows quickly build up a film. If you want clean, fresh air and less film buildup on your windows, put baking soda in all the ashtrays. Not only does the baking soda keep the air fresh, it makes putting out a cigarette less hazardous. The smoker doesn't have to reach and fumble to squash out the cigarette. Rather he or she simply sticks it in the baking soda. You'll find using baking soda especially helpful in the winter when you drive with the windows closed a lot. Every so often simply change the baking soda in the ashtrays.

Inside Wiper

You go out to your car in the morning and there's condensation on the inside of the window. You wipe it with your hanky and now have a wet hanky. You wipe it off with your hand and you now have a wet hand and streaks on the glass from the oil on your hands. The solution? Carry a blackboard eraser in your car. One wipe over the window with the eraser and the window is clear and your hand and your hanky are still dry.

Leak Check

The car is a few years old and when you checked the fluid levels last time it appeared you may have some leaks. Go to the grocery store and get a big cardboard box. Cut it so that it will lie flat under the car. Leave it under the car overnight and then the next morning pull it out to see if you have leaks, what is leaking and where it's coming from. Black spots mean oil, pink transmission fluid, greenish yellow coolant. Make sure you park the car on a level portion of your drive before doing this. And make sure the cardboard lies flat and doesn't get close to the hot catalytic converter underneath.

More Oil

You have to change a tire or simply have to rotate your tires and find getting the wheel cover off is the first chore, unloosening the lug nuts another. A tiny squirt of oil around the rim of the wheel cover will help when putting it back. Also give the lug nuts a squirt of oil and you'll find them much easier to loosen and then tighten the next time.

Touch Up

Unless you stand guard over your car in a parking lot, chances are someone is going to tag your door and leave a little scratch. You also will get a pock mark or two from debris kicked up from the road. Always keep a touch up pencil in the glove box to get a protective layer of paint over the exposed metal to prevent rust. If you don't have a touch up stick, a film of oil over the scratch will help until you can get one.

Paint Job

You finally broke down and had the car painted. You didn't like the original shade of red so you spent $600 and got a darker one. Fine, but whenever you get the car repainted make sure you get a small bottle or container of the same paint so you can perform minor touch ups later. My son purchased a 1976 Firebird that the previous owner had painted a maroon once used by Buick. Problem was that Buick dropped the color and there was no way to get a touch up. We had to settle for the closest color we could come by, which was three shades too light.

Ethanol

Ethanol is a blend of lead free gas and alcohol. On the gas station pump you may see a sign that says the fuel contains 10 percent alcohol. The reason petroleum companies began offering an alcohol blend was to stretch supplies of gas by diluting it with alcohol. Farm states were happy about the "gasohol" because the alcohol was made from corn. However, most new car owner manuals advise not to use more than a 10 percent blend. But you have no way of knowing if the concentration is 10 percent or 15 percent or more. Alcohol attracts water and except for the Stanley Steamer and a few others, cars were never made to burn water. Alcohol also has a tendency to eat rubber and plastic parts. Cars I've tested in which I used ethanol always had a tendency to obtain less mileage and be troublesome in cold weather starts and stalls. If you experience a mileage loss or start-stall problems check the gas you're using. Is it a diluted ethanol blend? Try a tankful or two of pure gasoline and see if that relieves some of your problems. Alcohol also was promoted as a cleansing agent. But where does the gunk go that is cleaned up? Into your fuel line filter. If you've been using ethanol, change the fuel filter and get in the habit of using pure gas again.

Injector Clog

If your car has a fuel injected engine, make note that several oil companies now have special additive fuels designed to help clean the fuel injectors. The opening to an injector is about the diameter of a human hair and the automakers found that deposits from lead free gas were clogging the injector nozzle holes and blocking the normal spray of fuel into the cylinders. By adding improved detergent additives to their fuel, the petroleum companies now boast they can get rid of injector clog after about two tanks of fuel. Mobil and Amoco both have special injector blend fuels and others are soon to be on the market.

Full Tank

More on fuel. The villain in cold weather starts often is moisture and condensation that forms in a half or near empty gas tank. Often motorists will pour special heat treatment additives into the gas tank in the winter to rid the system of water. The additives usually are alcohol that attracts water so it then can be burned out of the system. Keeping the gas tank full will reduce the likelihood of condensation buildup and the money spent on gallons of fuel will be more beneficial than spending it on several pints of additives.

Blow Dry Decals

You buy a car and it's got a dealer logo on it that takes up half the deck lid. Plus, the car is blue and the dealer logo decal is green and unsightly. If you want to rid the car of a dealer decal, bumper sticker or window decal, simply get out your hand held hair dryer, turn it on low, and make several passes over the decal until the stickum underneath is loose enough to peel off the decal.

Collectibles

Don't rule out a collector car when shopping for a used machine. No, I'm not talking about a Duesenburg, but perhaps a '66 Mustang. Dean Kruse, noted auctioneer with ITT International, auctions off thousands of antique, vintage, celebrity and assorted collector cars each year. "A car takes a jump in value on its 20th birthday," he told me. "Buy a 17 to 19 year old car, especially convertibles, and there's little way you'll lose and you could double or triple your money in a short time. I auctioned off a 1963 Corvette split window coupe for $44,000 that when it was 17 years old could have been bought for $5,000." Car shows are a good reference point to find out what's available and how prices currently are running. Car club members are a wealth of information. *Hemings Motor News* and *Old Cars Weekly* out of Iola, Wisconsin, are two other sources of collector news.

Alarm Systems

That car you bought sure looks great. Hate to have someone steal it. So you spend $600 for a system with whistles, alarms, bells and assorted other disabling devices. I once had a woman tell me she bought an alarm system to protect her $3990 Yugo. It mystifies me that a car owner will pay his or her insurance company for theft coverage which in the event of a theft will replace the car, and then will spend another $600 or more on alarms and bells. I've interviewed a variety of car thieves over the years. They all scoff at protection systems. If a professional theft ring wants your car, they'll get it — alarms, whistles and all. Have you ever been to a shopping mall and listened as the alarm goes off on a car? Did you rush off to call the police? Did anybody care the alarm was sounding or did everyone keep walking while casually mentioning: "Sure is loud, wish the idiot would come out and turn it off." That's the apathy you'll get with your alarm. A friend told me the story of the guy who spent $600 for a theft protection system, had the car stolen and the insurance company deducted the $600 from the settlement. It doesn't insure alarm systems. Keep in mind that thieves are consumers, too. When an alarm system is put on the market, who do you think buys it, takes it back to the garage, and finds the means of defeating it? Right, the professional thief.

One thief told me that people fake theft systems on their cars either with a non-working key lock on the fender or a series of decals in the windows. "When in doubt I walk up to it and kick the fender," the thief said. "If it's a fake, no alarm. If it isn't a fake, I can always come back and get it."

The comeback will be when the car is parked in the apartment or condo complex, the public lot in town where workers for the nearby office building park every day, the places that the particular car is parked on a regular basis.

But, you say, *your* theft system has a pager that will call you and let you know someone is messing with your car. If the thief hasn't gotten in and away with your car within 30 seconds he's a stumblebum. Probably the pager does work — if you're sitting on the fender of the car at the time.

La Erotica

You have your eye on that exotic European import and nothing anyone says will discourage you. If you do your own mechanical work the purchase of a limited edition car becomes less of a gamble. But if you are going to have to rely on dealership mechanics, can you afford their labor rates, much less their parts costs? If you want to save money on service and repairs by having an independent mechanic perform the work, you'd better call that friendly independent who has worked on your other cars to determine if he'll even work on this one. You may be surprised to find he won't touch that car — can't get the parts unless he himself goes through a dealer and has to pay top dollar, or simply finds it such a maze he doesn't bother.

LETTERS

Over the last 16 years at the *Chicago Tribune* I've received thousands of letters and phone calls from readers who either found themselves in trouble or who thought they were about to get in over their head and were seeking advice.

Usually the letter writer or caller starts out by saying, "I bet you never heard a problem like this before." Usually I have, but I wish I hadn't. I really don't know how Dear Abby does it.

The saddest part of fielding the letters and calls is that people write or phone *after* they've gotten themselves in a fix and now it's too late.

I vividly recall a man who called oozing with cockiness. He had purchased a car on Saturday and the salesman had used every dirty trick in the book. Every add-on profit fee known to man was tacked on the bill of sale. With each one the caller played dumb and agreed to the cost. Now it was a Monday and he called to tell me:

"Boy, I can't wait to see the salesman's face when I go back to the dealership today and rip up the bill of sale right in his face. I'm going to tell him I'm invoking my rights under the three day cooling off period and not buying the car. Boy, is he going to flip out," the caller laughed.

After the caller had relieved himself with laughter, I spoke.

"Sir, you mean you knew he was ripping you off and you let him do it anyway?" I asked.

"Yeah, he even charged me $250 for rustproofing a used car, and $150 for delivery and handling. He must have thought I was some jerk. The only reason I played along was to see the look on his face when I brought the car back after the cooling off period," the caller replied.

"Sir," I said, "There is no three day cooling off period on the purchase of a car. There's a cooling off period when you purchase an item from a salesman who comes to your door, but not on a car that you purchased at his place of business and signed a contract on. Sir, you're stuck with the car."

The man hung up.

This is but one example of getting into trouble, then seeking an out rather than seeking an out before you get into trouble.

Here are some examples of the type of mail and phone calls I get from readers. Please read them carefully and if you spot yourself being put in a similar situation, I hope you'll know what to do—or not do.

Some letters apply to used cars, some to new cars, some to the care and upkeep of cars. Others are included for your enjoyment.

Q. When I picked up my new $3990 Yugo, the dealer had a $48.70 documentation charge. I was told the fee is added to every car they sell. Please warn others about this fee.

A. I'm puzzled you're so worried about the documentation fee. On the bill of sale you enclosed, I noted you paid $289 for rustproofing, $239 for a rustproofing shield, $199 for a paint shield, $125 for a fabric shield, $49 for a sound shield, $489 for a "car saver system", $159 for a "stop theft" system, $348 for a burglar alarm and $199 for a security lock. That's $2096 in extras. Add the $48.70 documentary fee and it's $2144.70. In other words, you paid 54 percent of the value of the $3990 car for add-ons or a total of more than $6100 to take possession of a $3990 car. Yet you questioned the $48.70 documentation fee?

Q. I went to look at a Taurus and a Sable. The Sable had $900 worth of protection (rustproofing/fabric treatment, etc.) added and the Taurus a "dealer installed" sunroof for $1350. What's with the greed of these guys?

A. Law of supply and demand. They supply you a price and demand you pay it. Worst part, when dealers do this and people buy the cars, other dealers join in. But don't feel bad. A caller said when he questioned his Ford dealer's phony $95 documentation fee, he was told the money covered putting his warranty file on computer tape. When the customer questioned the equally phony $61 prep fee (how's that for an odd number?), he was told the money covered the cost of having a mechanic check the engine's timing.

Q. The dealer I went to has an obligatory rustproofing/upholstery treatment and extra dealer prep on his cars totaling more than $700. The salesman said, "We need our profit, too." I'm putting money in my clunker and waiting.

A. The only way this practice will stop is if more people do the same.

Q. I was about to buy a new Nissan Stanza when the salesman added a $100 documentation fee and second dealer prep fee of $150. He said all dealers do it. I said I wouldn't pay those fees, would look for a dealer who didn't charge them and walked out the door to my car. The salesman followed, said he would remove those charges and did. I saved $250.

A. Great. Hopefully more people will do the same.

Q. An ad in the paper said the dealer invoice would be next to the sticker price for a two day sale to show what a good deal you were getting. I went in that very night. No invoices. Next day, no invoices. A salesman said he heard rumors of such an ad but hadn't seen it.

A. Contact your local Federal Trade Commission office to complain about deceptive advertising. Also contact your local Better Business Bureau and the State Attorney General's Office, Bureau of Consumer Fraud. And send a letter to the editor of the paper. Contrary to popular belief, in major papers the editorial and advertising staffs don't associate. On the contrary, it's usually a hate/hate relationship. Someone in the ad department let an ad like that get in and let the dealer get away with it. Letting the editor of the paper know what the ad department is doing should help establish a policy of double checking the advertisers.

Q. I was taken by a dealer who made me sign a contract I didn't understand and he didn't give it to me for three days. Then he said I couldn't cancel. The price on the contract was different from the one he quoted me.

A. Unless the salesman was holding a gun on you or nailed your shoes to the floor, the instant you were confused was when you should have headed for the door. Don't even lift a pen until you know what you're getting into.

Q. I took my car to my dealership for a radiator drain and flush, dash lightbulb replacement, struts and timing check. The dealership called to say more work needed to be done— new brakes, new struts and shocks and a complete tune. I gave the okay and then checked around on prices and found I could get the work done more reasonably elsewhere. Too late, some work was done: $40 to replace the bulb, $60 for the flush and drain. I'm asking the factory to take the car back at book value.

A. You don't give the okay for work to be done *then* check prices, especially when an outfit charges $40 to change a bulb. Don't blame the factory. The dealership all but hit you in the head with a board to warn you to take the car and run and you didn't.

Q. I ordered a Ford Taurus station wagon and when it arrived five months later the salesman said my 1981 Toyota Celica trade-in had depreciated by $2000. When I signed the contract, he put on the bill of sale that the trade-in was "subject to reappraisal" but said not to worry. I was out of the market for five months while waiting and now don't have the car.

A. Some dealers play games with hard-to-get cars the way import dealers did at the height of the quota crunch when Japanese cars were so difficult to come by. It's greed. When a salesman puts "subject to reappraisal" on the bill of sale, you can count on it. He's low-balling you. You think the new car, for example, will cost you $10,000 with your trade in, which is $500 less than the dealer down the street had offered you. However, when the car arrives and is "reappraised" the dealer who offered the car for $500 less, gets the $500 back plus a lot more. Next time, when the salesman puts down "subject to reappraisal" hand him a typewritten document stating: "The dealer agrees to pay the buyer interest at the rate of six percent per month on the deposit between date of order and delivery." If he refuses to sign, tell him you refuse to a trade-in reappraisal.

Q. I bought a used '77 Mercury Cougar for $2500. The dealer assured me the car was perfect and not to do anything to it for a year. In a week the transmission cost $707 to repair.

In two months I replaced battery, idler arm, ball joints, U-joint, alternator and carburetor. Is it worthwhile keeping the car?

A. Be aware of the warning signs. A "perfect" car for $2500? Don't touch it for a year? To determine if the car is worth keeping you should do what you should have done in the first place. Find a good mechanic and have him check it out.

Q. I purchased an '83 Japanese import and have had occasion to look for repair parts — water pump, wiper blades, etc. Either the items weren't available at my local auto parts store or required a two week wait at the dealership, where the prices were high. Do the Japanese have a lock on all repair parts?

A. Collectively, the Japanese sell more than two million cars in the U.S. each year but among individual model leaders sales run about 150,000 units. That's not a large enough volume for all independent outlets to have a wide variety of inventory of replacement or repair parts. The lower the sales volume of a car the greater the likelihood you'll have to go back to the dealership for the part — and wait.

Q. I took my one year old Toyota in for service and told the dealer I wanted idle speed checked and set, brakes checked, body and engine mounts checked, nuts and bolts tightened. I was quoted $33, but the bill after a half hour was $72 because they also rotated the tires, changed oil and filter, aligned the wheels, flushed the coolant and cleaned the brakes. The usual consumer ripoff?

A. Body and engine mounts checked? Nuts and bolts tightened? The oil and filter change, tire rotation, alignment and coolant flush were the most logical services rendered. My concern is how all the work was performed in 30 minutes unless the mechanic was practicing for an Indy pit crew. You could complain to the Toyota zone office but when you tell them the dealer changed your oil when all you wanted was your bolts tightened, don't expect much sympathy.

Q. The dealer added a $25 fee on the used car I bought. Not until after asking three times and finally buying the car was I told the fee covered maintenance of the car on the used car lot. When I returned for service, they took the owner's manual out of the glove box and I can't get it back.

A. Perhaps you didn't pay your glove box fee. Don't pay for something unless you know what you are paying for. Maintenance of the car on the used car lot? In other words, he washed it. I once came across a dealer who was charging customers an $85 fee on each new car sale involving a trade-in. After months finally someone called and asked me to find out what the $85 was for. The dealer charged everyone buying a new car and trading in an old one $85, he said, to "prep" the trade-in for the used car lot. Absurd? Sure. Did he get away with it? Many times before we caught him.

Q. I visited two dealers charging interest rates of 6.9 percent and asked for a quote on a three year loan on $10,000. I was quoted $327 a month at one dealer, $330 a month at the other. Yet my credit union quoted me $326 a month for a three year loan on $10,000 at 9.9 percent rate. And all rates were APR (annualized percentage rate).

A. It pays to shop interest rates just like it pays to shop cars. To confuse you more, I called a major Chicago bank and asked for a quote on a three year, $10,000 car loan at 6.9 percent and was told the payments would be $308 a month.

Q. I went to a dealership, was quoted a price and then went to the finance office where they started working with a computer. A salesman said something about an extended service contract. I was given a big pile of papers to sign, which I didn't read carefully — I thought I was dealing with a company I could trust. The next morning, I read the papers and found I had purchased both an insurance and a service contract I didn't want. I decided to pay off the car three days later and they charged me a month's interest on the loan. I learned a lesson — ask questions and don't trust anyone. Believe it or not I have a degree in education, accounting and computer science. I felt dumb.

A. You may have misjudged the dealership personnel but your self-evaluation was on target.

Q. I received a postcard in the mail and it made me believe there was an emergency and my car had been recalled. It said "Urgent! Please call at your earliest convenience." It gave a toll free number to call. All they wanted was to sell me an extended warranty.

A. The same complaint comes in frequently and you're right, the notice makes it appear you have a serious problem when it cons you into calling to get a warranty sales pitch. Why the deception? It's no different than the postcard you get in the mail that starts, "You've already won a prize," except that rather than play on our greed for something free, it plays on our fear that something dangerous exists in our car.

Q. After purchasing a new car, I began getting mail urging me to purchase an extended warranty. I asked the dealer and wrote Detroit but no one would show me a copy of the contract so that I knew what I was getting. The dealer said the contract would be mailed after I signed the application. Why?

A. Because if you had a chance to read the fine print on what was excluded and what service and maintenance you had to perform, you probably wouldn't purchase the warranty. Then the dealer would be out a hefty profit. With no contract, how do you know the dealer won't pocket your money and that a warranty even exists? In your case, the factory offers a five year, 50,000 mile warranty on the car so why was the dealer trying to sell you an "extended" warranty?

Q. I bought a Ford Escort and paid $600 for a five year, 50,000 mile extended service plan. Checking the Ford district afterwards, I found there was no set price for the plan: dealers charge what the traffic will bear. I also later found that the coverage was for 36,000 miles, not 50,000 miles and feel the price was exorbitant.

A. You have to shop options just like you shop the cars. Do your homework. This is another reason I object to extended warranties to begin with. Your other problem is common and the reason you had trouble in the first place. You say you did your checking *afterwards*.

Q. I have an '81 Regal with 30,000 miles on it that starts well, idles beautifully and runs fine. Would it be feasible to just change the plugs rather than have a full tune?

A. If it's not broke, don't fix it. Thanks to electronic ignition and lead free gas, a plug change is all the tuning some cars need. When the plugs are pulled, examine any deposits (see chapter on plugs) for clues that other work might be needed.

Q. I'm annoyed. The body side moldings on my '81 Malibu came loose. A racket to create repair work for the dealer?

A. Detroit in the last few years has gone to adhesive type moldings to do away with drilling holes to fasten them on.

All the holes did was create a neat place to trap moisture for rust to form. Though some replacement moldings are expensive, they still are far cheaper than replacing rusted doors. The question here is, were the moldings installed at the factory or was the car delivered minus moldings and the dealer installed cheap ones on his own?

Q. I bought a Toyota Cressida and am very pleased but heard Japanese cars have thinner metal and dent more easily than U.S. made cars. Am I at greater risk in an accident?

A. Cressida sheet metal is 0.7 millimeters thick. By comparison, the sheet metal on an Olds Ciera is 0.8 millimeters thick. In either car, if the sheet metal were paper, you'd see through it.

Q. I placed an order at four dealerships to ensure getting a car. When the car finally arrived at one I went back to the others to get my deposits back but they're giving me a hard time.

A. Multiple ordering at four dealers to guarantee one will deliver is no better than some of the tricks and stunts salesmen pull. You don't deserve a deposit refund at those other three. If you want the order you place with a dealer to be firm and have him live up to it, you better be prepared to live up to your end of the deal, too. You gave each dealer a deposit to get you a car and that's a contract, friend.

Q. I went to a dealership to look at a specific car with a five-speed and found none. I made an appointment with the salesman to come back a few days later to drive one with a five-speed. I called first, told him I wanted to drive the five-speed, and he said: "Fine, come on out." I did and there was no five-speed. He said he didn't know when one would be available.

A. The salesman obviously thought once he got you in the door you would relent and take the automatic that was available. With no five-speed in stock, he wasn't going to make a sale and probably feared you would find one elsewhere. So he had nothing to lose by getting you in the door. Next time when you call ahead of time for a particular make or model, ask the salesman what color the car is that has the five-speed, does it have cloth seats, etc. If he says, "I don't know," tell him to look and call you back. It may save you the grief of a needless trip.

Q. I see cars with dealer plates on them at supermarkets, restaurants and nightclubs. Yet, I've never purchased a car with more than six to nine miles on the odometer. Tampering?

A. Cars used by dealers, staff, wives, kids, etc., are sold as demos. By the way, ever wonder how those cars you bought could be prepped properly and road tested to ensure there were no problems before delivery when they only had six to nine miles on them?

Q. Do you think $230 is too much for a paint stripe on a car?

A. Not if it's being promoted to colonel.

Q. Please comment on your views on synthetic oils. Enclosed is an ad for one at $6.25 a quart that you change only every 25,000 miles and filter every 12,500. I bought an '85 model car and don't want to experiment.

A. Are you trying to preserve your engine or avoid changing oil? You usually can find a quality motor oil on sale for 89 cents a quart and a name brand filter on special for $2 to $4. By changing oil and filter often, no more than every 3000 miles or two months of driving, you and your engine should enjoy lots of miles of motoring.

Q. Please compare the quality of all major brand motor oils.

A. Other than staying away from any oil not designated "SF" on the can, I've found the quality of major oil is directly related to the frequency with which it is changed.

Q. I purchased a Honda Civic station wagon in 1985 and had it back for service twice. It's sluggish, low thyroid. I'm heartsick. Is this what happens to an auto when it's made in the U.S.A.?

A. The car you bought isn't noted for speed — or having a thyroid. As for U.S. quality, the '85 Honda Civic wagon was built in Japan.

Q. I drove the wheel of my car into the corner of the curb near the pump at the gas station, causing the hubcap to pop off. I thought I had metal hubcaps. I discovered it was plastic but costs $50 to replace. Would the old-fashioned metal hubcap break so easily? Why must GM make things that don't seem so durable?

A. You ran into the island curb at a gas station and you are upset with GM durability? Anyway, the old-fashioned hubcap would have stayed in one piece — permanently dented. Whenever possible, lightweight plastic is substituted for metal in a car in order for the manufacturers to save precious ounces that translate into better fuel economy. Despite the ready availability of gas at relatively low prices, Detroit still has to contend with federal fuel economy laws on the books since 1978.

Q. Is it illegal to sell a car without shoulder belts? Is there any type or device that can be added to a car to prevent operation if belts aren't fastened?

A. About 1966, front and rear lap belts appeared in cars and in 1972 came separate lap and shoulder belts with a "put them on" warning buzzer. In 1974 the two became one system and interlock was added requiring belts to be fastened before the ignition worked (a Lee Iacocca idea, by the way). In 1975 interlock was abolished because so many people simply found ways to defeat the unpopular system. The carmakers have to have a belt system in all the new cars they sell, but it's not illegal for a dealer or private party to sell a used car without a belt system.

Q. Lap/shoulder belts are too difficult to fasten, muss clothes and are uncomfortable. Return to airplane type belts, please.

A. Have you ever seen a semi-trailer tailgate a 747 on the local interstate? Most people refuse to wear belts simply because they got in the habit of driving without them or because it's not considered macho to be seen with them on.

Q. My '66 Volvo needs leaded premium gasoline. A friend says she puts mothballs in her BMW tank, because the naphthalene in them boosts octane and prevents knock and afterburn. Will it hurt my car?

A. Once the laughter subsided at Volvo, BMW, the American Petroleum Institute and Amoco, all did some research and admitted that the naphthalene in mothballs will boost octane — but you may pay a stiff price. The naphthalene is soaked onto a holding agent, such as cotton, which doesn't dissolve in the tank. You could clog fuel lines and filters and in general harm the engine. And Volvo said it would take five percent mothballs by volume in your tank to raise the octane level.

Q. Why should I have more head room in my 1983 Delta 88 with no power seat than in my 1985 Delta 88 with a power seat? The only solution I see is to place two bowling balls on the front seat of my '85 Delta for a week or two in the hope they might depress the upholstery enough to give me the head room I need.

A. You answered your own question with the word "power." When you add power accessories to a car you typically lose room because of the controls and associated hardware needed to provide that option. The bowling ball routine is a bit outlandish. I suggest instead that you remove your shoes. Seriously, if two bowling balls placed on the seat for a week or two are enough to sag it by an inch or so, your complaint would be with Olds for providing seats that don't hold up to weight without giving in.

Q. I wasn't interested in the cars the dealer had on the lot but had a difficult time retrieving the key to my car which they had taken to test drive. When I suggested I would call the police my key quickly appeared.

A. Yours is a takeoff on a routine reportedly used by a former Chicago dealer who waited until a customer walked into the showroom and then parked a car behind the customer's to block his exit and give the salesman time to make the sale.

Q. The dealership I take my car to charges $5 for the use of shop rags. Creative?

A. So you be creative and bring your own rags and save the $5.

Q. While test driving a car I was stopped and ticketed for driving without a license plate. The dealer hadn't put one on. The ticket was for $50. I feel the responsibility of having a plate rests with the salesman.

A. So did the dealership which sent you a check for $50. Those taking a test drive of a new or used car should ensure that the same doesn't happen to them.

Q. As a warning to others, don't take delivery of a new car in the dark. By the time I finally signed the papers and took delivery, it was dark. I found the next day the car didn't have four-wheel drive as promised, no mud flaps, the wipers didn't work, the trunk was scratched, the rear bumper dented and scuffed. Now I'm stuck because I wasn't clever enough to walk out and come back the next day.

A. CAVEAT LOOKOUT. The Monroney label (window sticker) would have stated if the car was a four-wheel drive model, as would the bill of sale. How could you have missed spotting the absence of mud flaps, scratches, dents and scuffs immediately? There had to be some lights on at the dealership or in the lot. But you're right, you want to inspect the vehicle in the daylight before taking delivery. And you want all problems fixed before you take delivery.

Q. My wife drives about 2,000 miles a year and most trips are about 4 miles. GM says she should change oil once every 12 months. Comment?

A. GM doesn't say to change oil once a year. No manufacturer does. GM recommends 3,000 mile oil and filter changes for those whose travel is mostly the short haul, stop and go variety. Because your wife does so little driving, the 2,000 mile change is too long — her car must be experiencing moisture and condensation buildup from never warming properly during short hauls. I suggest a change every 3 months.

Q. When the odometer on my car reached 29,999 it rolled over to 40,000 instead of 30,000. The dealer says he'll replace the odometer but isn't sure what mileage to put on it.

A. The National Highway Traffic Safety Administration says the mileage on the replacement odometer must be the same as the reading on the odometer that's being replaced. So the new unit would have to be set at 40,000 miles. Sorry.

Q. I purchased a new car with a repainted hood. The salesman said the car had been damaged. Upon arrival home, closer inspection revealed that the door, fender and roof had been repainted and there were glass shards near the defroster ducts. Under provisions of the lemon law I returned the car for a new model. Also, the car had $995 in add-ons for paint, fabric, sound and vinyl shields and rustproofing. But the same package was a $495 option at another dealer.

A. A repainted hood and the salesman admitted damage yet you bought it anyway? Mistake No. 1. "Upon arrival home" you inspected the car and found more damage and broken glass. Mistake No. 2. Never take delivery before a thorough inspection and test drive. Though an unpleasant experience, you got a new car under the lemon law, which wasn't a bad deal. As for the add-ons, they were unnecessary at $495, much less $995. If you insist on running up the profit at your own expense, you at least should shop around for those services.

Q. I went to a dealership to check out used cars on a Sunday when it was closed. I spotted keys in one and, being honest, locked the car and took the keys with me. I returned on Monday with the keys. At first the manager was going to give me a car out of gratitude, then he said he'd give me a good deal. I told him I only had $5,000 to spend. I asked about an '86 Olds and was told $9,995, but for me it would only be $8,500. He finally came down to $8,100 and with the extended warranty and documentary fee it was mine for $8,910. He said in order to talk to the finance man I had to put up a $100 deposit. To make a long story short, the next day I decided not to buy the car but they said they had already cashed my check and I couldn't get my money back for 12 days. They had no business asking for my money in the first place.

A. Your first mistake was in taking the keys and it went downhill after that. With $5,000 to spend why'd you even look at a $9,995 car? Why did you give the man $100 to determine what your payments would be? Do you have to give the butcher $100 at the market before he slices up a steak to your liking? They had no right asking for the $100, you had no business giving it to them. And you don't need the extended warranty and certainly shouldn't pay a documentary fee.

Q. I object that Ford Motor Company provided my name and address to a dealership in New Jersey though I bought my car in Chicago. An official looking postcard arrived that asked me to call an 800 number and have my owner's card in hand when I did. I assumed a problem with my car. Once on the phone I was subjected to a high pressure sales pitch for an extended warranty for $500. The supervisor admitted an extended warranty could be purchased through any dealer but said this dealership was sending cards as a service to Ford customers.

A. The solicitation wasn't from Ford. When Ford or GM or Chrysler or Toyota or Nissan or Honda solicit extended warranties they're up front about it and make it clear in their mailings. In your case, an independent outfit selling extended warranties obtained lists of customers who hadn't purchased them when they bought their new cars. Many motorists complain about this practice because the postcards lead them to believe there is an urgent problem. Treat the cards as you would the notice in the mail: "You've just won a free . . ."

Q. I called a Volvo dealership and asked the price of an antifreeze change. I was told $85, but "more if something wrong is found." They said Volvo antifreeze costs $10 a gallon and that I shouldn't use any other antifreeze in my car because when mixed with North American water the non-Volvo antifreeze will hurt my radiator. Are they just being cute with me?

A. Cute isn't the word. And when they say "more if something wrong is found" suggests they'll be looking for something wrong, don't you think? The only antifreeze that shouldn't be mixed with North American water is Canadian Club.

Q. I purchased a car and the business manager automatically included optional credit life insurance on the contract without asking me, saying 99 percent of his customers took it. I asked what the insurance premiums added to my monthly payments and he said only $7 to $10 a month. I refused the policy and it reduced my monthly payments by $23.44. Someone must be making a bundle on these policies.

A. The policies aren't given away. You were wise to question the charge. You didn't state how long the contract was for, but if for only 36 months, you saved $843.84.

Q. I own a GM car. When I start out on a cold morning the power steering is finicky and sluggish. What gives?

A. GM has a problem called "morning sickness," a malady in which it takes a couple of minutes for the power steering to work on cold mornings. Those experiencing the problem have been owners of 1982–1984 Buick Skyhawk, Skylark, and Century; Cadillac Cimarron; Chevrolet Cavalier, Citation, and Celebrity; Olds Firenza, Omega, and Cutlass Ciera; and Pontiac Sunbird, Phoenix, and 6000 models. Also, owners of 1985 Buick Electra, Cadillac DeVille, and Olds 98 cars. Owners of other GM cars for other model years also have complained of the problem, but GM has agreed to make free repairs to those with the problem in the cars named above or reimburse owners who already paid to have the work done. There is, however, a 5 year/50,000 mile restriction to the free repair or reimbursement. Call your local zone office and dealership for more information.

Q. My home was flooded and several cases of oil were under water for some time. A number of them were paper containers and the sides are now soft. Has the oil been contaminated?

A. Officials at the American Petroleum Institute said any water in the cans is probably only a minor amount. The greater fear is if you spot rust marks on the can tops. The rust may have seeped into the oil. If you spot rust, don't use the oil.

Q. I took my car to a service store for an oil change. I told the manager to also check the front end, which was making a noise since a recent accident. He drove the car and said it was the constant velocity shaft and outer bearings—$350 plus labor. To avoid the hassle of going elsewhere, I told him to go ahead with the work. He called and said it would take $268 more after actually looking inside. The final bill was $933.80. I paid by check, then stopped payment. They turned around and dunned me for the bill, including freezing my bank account.

A. When you were quoted the problem and the price to fix it based on a ride, not having the mechanic actually look inside, you were asking for grief. You should have gotten a second opinion before agreeing to any work. Based on knowing what the problem was after getting a second or third opinion, you then could have shopped for the best price. By approving the repairs at the price quoted, you had an obligation to pay and not stop payment. You didn't avoid any hassle. You had a mighty expensive oil change.

Q. Now that Pontiac has discontinued the Fiero, will it become a collector car?

A. Maybe yes, maybe no. But Fiero and potential collectible cars deserve a separate chapter of their own. So read on about the problems from callers and then I'll talk about collectibles.

Then there were the callers.

— A woman called saying the salesman admitted making a $65 error on the bill of sale. But he refused to subtract the $65 mistake in addition even though he had a $350 "saver package" built into the bill of sale that included high-profit rustproofing, interior fabric treatment and exterior lusterizer. Add all numbers before signing, not afterwards.

— Another woman said the salesman bumped up the price of her car by $250 because when it arrived from the factory it was already rustproofed. Oddly enough, the car carried a decal on the window from a franchise rustproofing outfit. Even odder is the fact no factory adds rustproofing material. They use galvanized metals and zincrometals but no manufacturer drills holes and puts in rustproofing goop at the factory.

—Then there was the man who called to complain that he bought a new Chevy when an 8.9 percent financing rate was in effect, but the car arrived when a lower 7.9 percent rate was being offered. Surprisingly, his complaint wasn't about the one percent difference. Rather: "After I got home I looked at the papers and instead of 8.9 percent my rate was 23.9 percent." Before signing any papers double check the little line that tells you what the financing rate will be. If the line isn't filled in, don't sign. And don't wait until you get home to read those papers.

—I once casually mentioned in a column that car buyers shouldn't use their dealer's water closet facility or they probably would end up paying a Latrine Tax. Well, a Ford dealer came close—with a Profit Tax. Yes, a profit tax. A reader called to say the dealer told him that the dealership had received a letter from Ford informing it to begin collecting a tax on the profit it makes on the sale of cars in addition to the county sales tax owed. In the reader's case, he would be charged a $37.50 Dealer Profit Tax.

I called the local district sales manager for Ford who said he was unaware that such a tax existed, much less that a letter existed telling dealers to collect it. I called Ford in Dearborn, and all I got was laughter. There was no tax and no letter instructing dealers to collect the tax. The dealer was engaging in deception. Call your local state attorney general's office.

—A woman called on behalf of her sister Maria, who doesn't speak English. Sis said Maria saw a 1979 Lincoln she liked for $20,000, talked to the Spanish-speaking salesman, and was told a deposit would hold the car until her credit was checked. If her credit wasn't approved, the dealership would refund her $1000 deposit. Oh, and sign here please. The credit wasn't approved and Sis got a letter from Ford Motor Credit Company, the auto maker's financing subsidiary, telling her so. Maria went back to get her deposit. That's when the salesman showed her the paper she signed stating deposits weren't refundable. Maria was out a car and a deposit.

I called the dealership and got Junior, the used car sales manager. Junior said Maria didn't want a '79 Lincoln, but an '83. In fact, Maria came in several times, each time changing her mind on what car she wanted. Maria's credit was approved, Junior said, but she backed out of the deal. And yes, the dealer will keep the $1000 deposit.

"Our people spent lots of hours, invested a lot of time in her," Junior said. "She was here umpteen times over two weeks. Besides, the salesman spoke Spanish so she can't say she didn't understand. He showed her in black and white that the contract says no refunds of deposits. We never make refunds. If she wants to apply the $1,000 toward another car, fine, but no refund."

In talking again with Maria's sister she mentioned the $20,000 car wasn't for Maria but for Maria's 20 year old son who comes home to Chicago once a month from a managerial job in Houston. Maria wanted out of the deal because, without credit, she couldn't afford the car and it only would be repossessed.

Comment: When a 20 year old with a good job hits on mom for a $20,000 car, mom should hit on son with a lug wrench. The money he spends in air fare to come home once a month could pay for a car. Besides, Houston is knee deep in used luxury cars and he could have bought one there the old-fashioned way — with his own money.

Maria should have her head examined for walking into a dealership without a command of the language and no one at her side; for buying a $20,000 car with only $1,000 to her name; for handing over $1,000 to begin with; and for not getting the refund promise in writing.

As for good old "we never make refunds" Junior, for

starters in checking the book value of a used '83 Lincoln at the time, prices ranged from $13,000 for a Town Car to $18,000 for a Mark VI, meaning Junior had leeway built into the asking price to make a profit without keeping the deposit. As for the time spent, that's how cars are sold, by salesmen investing some time with the customer. If customers are going to be required to pay for the salesman's time they should be given a stopwatch when they enter the lot. It appears Junior has found a way to make money the easy way—by not giving it back.

COLLECTIBLES

Bad buys go good

Mike Losh, Pontiac general manager, had barely gotten the words out of his mouth ("We are going to discontinue the Fiero at the end of the 1988 model run") when the phone started ringing.

"Will the Fiero become a collector's item?" the first of several callers asked.

My initial impression was no. The two-seater car wasn't a very good car to begin with, which is one reason it's on my list of used cars not to buy back in the chapter on "Cars to Look Out For—Lemon Tarts."

The car is small, cramped, underpowered with the 2.5 liter four cylinder engine, boasts of mediocre ride and handling, and is saddled with the fact that for too long you could buy any Fiero you wanted as long as it was red.

The 2.5 liter four cylinder was designed for high mileage, not high power. The intent of the car in the first place was to provide GM with a high mileage economy car that looked sporty. Fiero was developed when gas prices were skyrocketing. GM decided to come up with a stylish economy car. By the time the car arrived on the market gas prices had retreated and consumers were looking for cars with low 0 to 60 m.p.h. times, not necessarily high 0 to 20 m.p.g. ratings.

In Fiero's favor is the fact you can't dispute that the two-seater represented a dramatic design departure for staid and conservative General Motors when it was first brought out in the 1984 model year. And you have to appreciate the flexible plastic body that won't rust.

While the regular Fiero leaves a lot to be desired, you can't fault the GT sporty version with its 2.8 liter V-6 engine. The Fiero GT has very sporty styling going for it. It's cuter than the regular Fiero.

I've talked to some collectors, including Greg Grams. Along with his brother Bill, Greg runs the Volo Antique Auto Museum in the far northwestern suburbs of Chicago. The brothers are members of the Classic Car Club of America and just about any other group dedicated to the preservation of automobiles.

Grams insists Fiero could be worth money someday and should be held onto. But what collectors of the future will look for is the top of the line GT sports version with the 2.8 liter V-6 and 5-speed manual. Plus, he warns, make sure the car is loaded with all the bells and whistles from air to AM/FM.

Personally, however, since the 5-speed is arthritic, I'd still prefer the automatic.

As for the base Fiero with the 2.5 liter engine and the original 4-speed manual? Forget it.

How long will it take for the Fiero to be worth anything as a collectible? Grams insists five years — minimum. I'd lean toward ten.

Fiero brings up an interesting point. What about cars available today that could be worth some dough as collectibles of tomorrow?

I came up with a list for you to consider and perhaps argue over:

1. *Fiero GT* — Sporty looks and with the 2.8 liter V-6 a decent performer when it first came out in January 1986. With the plastic body it will have staying power.

2. *Cadillac Seville/Eldorado, Olds Toronado, Buick Riviera* — 1980 to 1985 models with the distinctive styl-

ing, huge size, lots of weight, big V-8 engines. The downsized replacements in 1986 were so sterile that GM will replace them all between 1989 and 1990. In the meantime, these old models have become hot items and are in demand, especially the so-called bustle-back Seville. In some cases an $8,000 '83–'84 Seville is now commanding $15,000. Propping up values more is the fact that when you subtract those powered with the infamous 350 cubic inch diesel engine, you don't have many left to choose from.

3. *Buick Regal Grand National GNX* — The 1987 model was an extremely limited edition number. Only 542 copies were built with its lightning fast 5 liter, fuel injected, turbocharged engine with a 0 to 50 m.p.h. time of less than the blink of an eye. Initial price was about $29,000. Some who got ahold of the car are asking $40,000 to $60,000. They don't — and won't — build 'em like the Grand National GNX any more.

4. *Cadillac Allante* — The luxury front wheel drive two-seater first appeared in the 1987 model year. It was a cooperative venture with Pininfarina of Italy. Worth the $50,000 plus price tag? No! A terribly overpriced rival to the Mercedes-Benz 560SL two-seater it's supposed to compete against. But only 2,517 of those 1987s were built and sold. Low numbers of an all-new model whose body was designed by Pininfarina of Italy should be in demand in the future.

5. *Buick Reatta* — Another two-seater. This one was priced at less than half that of the Allante, only $25,000. Good looks, very good engine, everything the Allante should have been for half the price. Again, limited availability — only 3,000 cars in 1988, its first year — and reports of some dealers getting $3,500 over sticker for the hard to find models. A keeper that bears holding onto for appreciation.

6. *Chrysler TC convertible by Maserati*—Not often you have a car shown in prototype form for 3 years prior to introduction in the summer of '88 as an '89 model, but Maserati took its time bringing out this luxury two-seater. Front wheel drive, turbocharged, and again very limited numbers. Full year production capacity was less than 5,000 units and with midyear intro, well, obviously the "first off" models will be rare.

7. *Chrysler Imperial*—1981–1983. Okay, it's on the list of lemon tarts because it's little more than a bustle-back version of the old Chrysler Cordoba. But styling was distinctive, the model run was brief, sales were paltry, and there aren't that many around.

8. *Cadillac 60 Special*—1987–1988. An experiment by Cadillac to answer its critics who complained its cars have gotten too small with downsizing. The 60 Special looks like a baby limo. Basically it's a 5-inch stretch of the sedan DeVille. Price tag about $34,000. Only 1,000 built each year.

9. *Mustang GT*—Any car, any color, any year (except 1974–1978—see chapter on Lemon Tarts) as long as it's got the V-8 engine. The dramatic rise in Ford's fortunes in the last few years is having an effect on the renewed popularity of its older cars. And everyone wants to collect sports models.

10. *Camaro/Firebird*—Any IROC, any TransAm. Perhaps the worst ride and handling cars in the world. Walk a mile in any man's shoes, but driving a mile in IROC or TransAm can do you bodily harm. Yet, can't beat styling or the fact teenagers will vow sainthood to get their hands on one. If insurance rates for youths wanting these cars keep skyrocketing, you could find performance dramatically toned (and tuned) down, which would

make older and current models even more attractive and in demand.

Also worth noting are any Lincoln Town Car, Mark and pre-1988 rear drive Continental, which should enjoy popularity equal to that of the 1980–1985 Seville/Toronado/Riviera from distinctive styling and V-8 engines. The Toyota Mr-2 two-seater may share the same fate as Fiero as insurance rates soar.

Those cars I'd pass on would be the Merkur XR4Ti sold through Lincoln-Mercury, especially with 5-speed; the downsized and terribly bland 1986 Seville, Eldorado, Toronado, Riviera; and any Cadillac Cimarron, finally discontinued (along with the Olds Firenza subcompact J-car) after the 1988 model year. While Cimarron had low volume, it didn't have any distinctiveness in styling to make it a memorable or desirable collector piece.

A final note to those who think the car in the barn is worth a ton of dough, or those who come across an old hunk sitting outside a barn and think they'd be wise to pay a ton of dough to take possession.

Old doesn't mean valuable. Auctioneer Dean Kruse of Kruse International passed along some handy advice to me once about circumstances that should make you cautious about the potential collectibility of a car.

I pass these tips along to you. Think twice if:

The car has been stored outside and subjected to the elements (and varmints).

The car has rust spots on it. A little rust is a lot of rust.

There's no original authentication of ownership history.

It has a V-6 engine when a V-8 had been offered. The V-8 commands the most value.

It wasn't popular when new. If people didn't like it then, they won't pay any more money for it now.

It has an amateur restoration that either was done at

home or by the lowest bidder with parts that "came close" to the original.

It has been poorly customized using bargain basement parts.

It has been wrecked at least once.

It's purple or mustard green.

It has more than 30,000 miles on it. A 1955 T-Bird with 120,000 miles loses some of its appeal.

GOOD BUYS, BAD BUYS

What's new?

The Good, the Bad, and the Ugly. No, not the Clint Eastwood western but the description of vehicles available to you in the used car market.

To help narrow the selections, to recognize and avoid the bad and ugly, here's a rundown on the vehicles I think you should strongly consider when checking out the merchandise on the used car lot.

But, what the heck, why not start with the failures first, the ones you should quickly slip past on the way to the prime cuts.

1. Any turbo-charged car. A turbo typically is a crutch, a means of injecting some life into what is a rather power-starved engine without actually putting a more powerful engine in the car. If the turbo needs to be fixed, the expense is costly.

2. Any GM car with the 350 cubic inch V-8 diesel engine. A gas engine converted to diesel. Nothing but trouble and costly repairs.

3. Any used Vega or Pinto. When new both were beaters.

4. The 1981 Cadillac Eldorado, DeVille, Seville and Fleetwood with the V-8-6-4 engine. Variable displacement from one engine for power or economy. Good idea, but miserable execution.

5. Any car still left standing with an AMC or Renault or Peugeot or Sterling or Yugo nameplate on it.

6. Any Chrysler vehicle with a Mitsubishi 2.6 liter four cylinder engine under the hood. Also used in the 1984 Dodge Caravan and Plymouth Voyager. Didn't run well, or for long.

7. Any Ford Bronco II, the compact sport utility vehicle with such a short wheelbase you felt tipsy going into a turn; or any Jeep Wrangler, a vehicle so crude that even Mother Theresa would find fault with it. Unzip a window to pay a toll???

8. Any Audi 5000, the '78-'86 models branded with charges of sudden acceleration. Government safety agencies concluded after a lengthy investigation that the reason for the acceleration is that drivers pressed the gas pedal rather than the brake pedal, but the damage to the image was done. My complaint with the cars is that sudden or ordinary old planned acceleration aside, most any Audi prior to 1990 wasn't much of a car.

9. Any 1986–1988 Buick Riviera or Olds Toronado, the victims of GM's downsizing campaign to shrink exteriors and make its cars more fuel efficient. In downsizing these cars GM sacrificed character. Bland machines. To undo a wrong, GM brought out a totally redesigned Riviera and a totally new vehicle called the Olds Aurora sedan to fill the Toronado void in 1995.

10. Cimarron, the 1985–1988 Cadillac imposter, a Chevy Cavalier wearing a leather mask to fool folks into thinking it was really a luxury car.

11. Any 1985–1989 Dodge Diplomat or Plymouth Grand Fury — cop cars.

12. Any 1988–1993 Ford Festiva or its 1994 to date replacement, the Ford Aspire. Built in South Korea. Very third world.

13. Any 1989 to date Ford Probe. Originally planned as Mustang replacement. Bland styling matched by bland performance.

14. Any car named H-Y-U-N-D-A-I E-X-C-E-L. Replaced by Hyundai Accent for 1995. Dealers didn't even want to take them in trade — Hyundai dealers.

15. Chevy Lumina/Pontiac Trans Sport/Olds Silhouette mini vans. Great idea — front wheel drive mini vans that went Chrysler mini vans one better by offering never-rust plastic body panels. But the execution was feeble. Long, sharply slanted front ends didn't look too bad from the outside, but gave you the impression from the inside that you had a couple of snowplow blades attached to the front bumper. Black dash padding in later years helped camouflage the length of that snout, but plans were already made to bring out more appealing metal body versions for 1997.

16. The 1984 Pontiac Fiero, first year introduced, and most any Fiero thereafter with a sadly underpowered four cylinder engine and balky manual transmission — and every Fiero with a four and manual was underpowered and balky.

    Now some might say, hey, wait a minute, what about Fiero collectible value as a piece of Americana? The rule in collecting is that any car that's not very good when new doesn't get any better with age.

    So even if you are collector-minded and feel that if you find an old car and restore it, it may be of value someday, all you'll be doing is throwing good money after bad if you purchase a Fiero or:

17. Any '74–'78 Ford Mustang, the downsized bulbous models with meek four cylinder engines as the industry focused on fuel economy and abandoned performance.

18. Any 1981–1983 Chrysler Imperial, which was merely a Cordoba with a bustle back deck lid added; or any Ford EXP or Mercury LN7, Ford's 1982–1985 experiment with two seaters that proved to be styling disasters.

The pretenders out of the way, let's turn to the contenders, the vehicles I advise checking out, the ones that stand out from the hundreds of nameplates you'll find on the used car lots.

Toyota Corolla—Underwent a design change in 1988 and again in 1993 when a driver's side air bag was added. Dual bags standard in 1994. Maybe not the best looking vehicle on the market, but you sure don't see a lot of them lined up outside the repair shop.

Chevy GEO Prizm—The Corolla with a Chevy nameplate. Usually priced far less than the Corolla since most people don't understand it's a Toyota underneath.

Saturn—Those first models in 1991 were noisy with automatic transmission, but with each succeeding model year the commotion quieted down. Very good value for the money. Plastic body panels won't rust. Excellent fuel economy. Good interior room. Driver's bag added in 1992, dual bags in 1995. Traction control offered with automatic transmissions since the 1993 model year. Great snow belt option. Coupes provide low cost—and lower insurance premium alternative—to Camaro/Firebird. Newly instituted Saturn program allows you to return any used Saturn within three days of purchase for full money-back refund with no questions asked — or if you change your mind after 30 days/1,500 miles, you can return it for another car. Plus, you get a warranty and 24-hour roadside assistance.

Mazda Miata — A gem. Okay, it's small and cramped and hardly screams when you press the pedal to the firewall, but the little drop top roadster exudes charm and character and fun from every pore. A real "triumph" — and the intent to embarrass the British automaker is intentional.

Mercury Villager/Nissan Quest — Same vehicle, different names. Designed by Nissan, built by Ford, the main attractions are that they ride and handle and are capable of being parked like cars even though they are mini vans, and the third seat in back slides forward to increase cargo hold without having to be removed, a sometimes cumbersome and back-breaking chore in most mini vans.

Chevy Cavalier — Noisy, cramped, underpowered, but a ton of them around so prices should be respectable for a knock-around vehicle during the summer, for school, or to get to the train station. The all new 1995 version is quiet, roomy, peppier, and has dual air bags. When sufficient quantities arrive on used car lots in a few years, the 1985–1994 versions will loose their appeal.

Ford Mustang — The 1994–1995 styling remake, with dual air bags and the performance, ride and handling, and quiet, has what just about all of its predecessors were lacking.

Honda Accord — Up until the 1990 model year when they were redesigned, enlarged and overpriced. Those early 1985–1989 models were the benchmarks for compact cars until Honda tinkered with the dimensions.

Toyota Camry — Guess what vehicle the Energizer Bunny travels in. Usually trouble and maintenance free. Keep changing the oil and filter, and this is a vehicle that gets passed down through generations of drivers in the same family. Prior to '88 a V-6 wasn't available. The V-6 provides more pep and quieter operation. Driver bag added in 1992, dual bags in 1994. Prices are high and the salesman will say, "But it's a Camry" to which you should reply, "But it's MY money."

Ford Taurus/Mercury Sable—First came out as 1986 models, but the best of the lot are 1992 and newer. Driver's bag in 1990, passenger bag in 1992 (both standard in 1994). The 3.8 liter V-6, added in 1988, is the preferred engine for optimum performance [other than the 220 h.p. V-6 in the SHO (Super High Output), of course]. Some early models had a tendency to feel heavy in the wheel. Lots of cars to choose from, especially Taurus, since it reigned as the industry sales leader in 1992 through 1994 (and at last look is still going strong). Careful though, since lots of these cars were sold to fleets and rental companies and could have wear and tear not evident if you only shop for a car with a good wax job and vacuumed carpets.

Buick Century/Olds Cierra—Cousins. Plain jane, nothing fancy or frilly or very technologically advanced sedans that didn't get a driver's side air bag until 1993 (optional, standard in 1994). But since they are simple machines, they are easy to fix and not too costly to repair. Excellent choice for older motorists looking for basic transportation and decent room without going all the way up to a Park Avenue or 98. The 3.3 liter V-6 added in 1989 is the preferred engine (replaced by a 3.1 liter V-6 in 1994). Take a pass on any four cylinder.

Olds Cutlass Supreme/Buick Regal—1990 and newer when the mediocre coupe was joined by a sedan. Go the sedan route, not the coupe unless the car is going to be a runaround for high school age youth who scorn sedans. Avoid the 2.8 liter V-6 (1988–1989) in favor of a 3.1 liter V-6 (1989–1995), while the 3.4 liter V-6 is best of all for performance (1991–1995). Decent enough looks to attract youth and first-time drivers who need more size and weight wrapped around them than a Civic or Cavalier. ABS optional since 1989, standard since 1994. Driver's bag in 1994, passenger added in 1995.

Pontiac Grand Prix—Okay, same platform as the Cutlass Supreme and Regal, but the best car in that GM-10 or W-body trio. Good choice for youthful drivers by giving them some size and weight in a stylish package (again—sedan, not coupe). Better ride and handling than its Olds/Buick cousins with the SE sports suspension.

Our favorite Pontiac and one of our favorite models new or used is the Grand Am. Excellent styling, above average mileage, a car that attracts luxury, economy and sports minded buyers. The 1985–1991 models not as peppy as later 1992 and newer versions, which brought back the V-6 engine previously only offered through 1987. There was some playing with turbos to boost power early on, but avoid the turbo. Lots of Quad Four, 16 valve four cylinder engines in the 1992 and newer models, but the 3.3 liter V-6 (1992–1993) and 3.1 liter V-6 (1994–1995) are preferred for smooth, quiet operation. GT versions very dressy. Driver bag added in 1994.

Buick LeSabre, Pontiac Bonneville—LeSabre for the older buyer concerned with luxury, Pontiac for the younger buyer focusing on performance. Bonneville SSE prior to redesign in 1992 when rear end became bloated is especially nice. However, it was 1992 when a driver's bag was added. SSE was one of early GM cars with ABS (optional 1988, standard 1993, passenger bag added in 1994). LeSabre offers just about everything as larger and more expensive Park Avenue—except the price. Warning—leather seats nice, but if you live in the snow belt, you might get a cold butt from it and in the sun belt you'll get a toasted butt out of it.

Chevy Caprice—Any car (except carbureted), any model year prior to 1991 when it was redesigned and became the darling of the taxi fleets and police departments. Prior to 1991, this car is old reliable. Anyone can fix it and parts are readily available. Plenty of room and ample power. Rear drive so you can tow.

Ford Crown Victoria—Unlike Caprice, the better models are post-1992 versions when they underwent a design change and kept the room and comfort but wrapped it in a more stylish package. Another rear wheel drive tow vehicle. Fuel economy is the only problem. This is not a gas miser.

Cadillac Seville—The 1992 and new versions, especially the STS with a suspension system normally reserved for the top of the line Pontiac performance model. In '93 a 32 valve North Star V-8 arrived that idled so quietly Cadillac added a starter interrupt feature that would disengage the starter if you turned the ignition key back to "on" while the engine was idling.

Dodge Caravan/Plymouth Voyager/Chrysler Town & Country mini vans—Pioneers in a new market segment introduced by Lee Iacocca. Best looking, best performing of all the competition that eventually entered the field. The mini vans first arrived in 1984 but the 1991 and newer models are the choice. Avoid at all costs the 2.6 liter Mitsubishi four cylinder engine offered in 1985-1987. The 3.3 liter V-6 offered in 1990-1995 and 3.8 liter V-6 in 1994-1995 are the preferred choices. Driver's bag in 1992, passenger bag in 1994.

Dodge Ram truck—Total remake in 1994 and an extended cab version with a seat in back for passengers added for 1995. Gas or diesel engines for high performance or high mileage. Massive interior space with a fold out from the seat office table, and a slew of racks or storage bins along the rear wall. The vehicle that reawakened truck loyalists to Dodge.

Chevy GEO Storm—Now you see it (1990), now you don't (1993). Sporty looking economy car, a fun machine with good interior room, fairly peppy and fuel efficient engine. Driver's side air bag. Died off because the sharp increase in value of the Japanese yen against the U.S. dollar made it too costly to import from Isuzu, GM's Japanese supplier. I was privy to some prototypes of future Storm models just before the decision to cancel the program. The styling of those Storm wannabees but neverwillbees was sensational and put Camaro sheetmetal to shame.

Now let's look ahead a bit and do some forecasting about which of today's new 1995 models may become choice selections on the used car lot a few years from now.

Here goes:

Olds Aurora—Excellent styling, a car that signifies that Olds finally got its act together. A 4-liter version of Cadillac's North Star V-8 under the hood. Limited supplies and very high demand have kept inventories low. At $32,000 to $33,000 when new, this car has just about everything the $40,000 to $50,000 Japanese luxury cars have—except the high price tag.

Chevy Cavalier/Pontiac Sunfire—After an eternity the Chevy/Pontiac subcompacts have been redesigned. Roomier, quieter, and more lively than their predecessors (Sunfire replaces the old Sunbird). Dual air bags standard.

Chevy GEO Metro sedan—Because it's so small, safety has always been our concern with this car, but for '95 it adds dual air bags and anti-lock brakes and increased interior dimensions to give you some elbow room. At quick glance it looks like a Cavalier.

Buick Riviera—Nice redesign of an old nameplate. Only a coupe and only a V-6, but attractive aerodynamic styling that has character.

Ford Contour/Mercury Mystique—Contour the primary choice because I don't like the front end styling on the Mystique. With top of the line suspension and V-6 engine you have what acts like a lower cost version of the larger Taurus. The media has criticized rear seat room, but most of them forget Contour isn't a replacement for a midsize Taurus but rather a replacement for the old compact Ford Tempo, which didn't have much room either.

Chrysler Cirrus—Contour's compact rival. A styling

masterpiece. One problem in fitting all that car into a compact package was that there was no room left under the hood for the battery, so it had to be located under the front fender. Another gripe—no Chrysler-built V-6 engine.

Chevy Blazer—The 1995 remake eliminates most of the problems of the older models—there's more room, more carlike ride and handling, less wind noise from the aerodynamic curves of the sheetmetal, along with the addition of a driver's side air bag. The shortcoming—no passenger side bag until 1997.

Ford Explorer—Less boxy and truck-like than previous models and you'll love the new dial-it-up four-wheel drive. Turn the button to 4WD and let it go to work on its own. If the computer sensors detect wheel slippage, four wheel drive is automatically engaged; if wheels aren't slipping, you run in two-wheel drive and save on fuel. Drawback—no engine upgrade as yet for a needed boost in power.

Dodge/Plymouth Neon—High mileage commuter cars with spacious interiors, especially in the back seat. Dual air bags standard. The drawback—initially only a three-speed automatic.

Chevy Corvette ZR1 or King of the Hill. Last models (448) built in 1995 and all are instant collectibles. A high performance successor not due out until 1997.

QUESTIONS AND PROCEDURES IN BUYING A CAR

Here is a series of questions you should ask and procedures you should follow in purchasing a used car. Following them and insisting on the answers before signing any contract can save you a lot of time and money.

1. What do you want and what do you need from the car? Think this through. The red Camaro will look very good on the lot but it won't hold the family.

2. What is the budget limitation? If you have $5000 to spend, then you must limit myself to what's available at that price. Don't play the monthly payment game the salesman will want to get you into. That's the old "only $300 a month" routine. You must understand that $300 a month doesn't sound bad, but $300 a month over 36 months is $10,800. A salesman will want to work up or down from a specific monthly payment figure. You end up talking in terms of $200 or $250 and the $50 spread doesn't sound bad. But every $50 difference over 36 months is $1800. If you can afford $5000 for a car, that's about $138 in monthly payments over 36 months.

3. Where will you obtain financing or will you pay cash? New car financing is relatively inexpensive with all the discount programs the manufacturers offer. Don't be fooled into thinking used car rates

are running about the same as new car rates. You can pay double or more. Check your local bank or credit union *before* you start looking for a car to (1) find out if you qualify for a loan and (2) find out how much of a loan you will qualify for. Sure, the credit union official says in the hallway at work that you'll qualify for a loan, but later you find the loan policy is only up to $3000 for used cars. And most important is (3) what is the rate of interest being charged? The rate will vary if it's a 24 month or 36 month loan (you won't find many places eager to give you a 48 month loan on a five year old car). If you have to have a car and find the only loan you can get is at 16 percent, but you have $2000 socked away in a savings account paying six percent interest, you might want to use that $2000 and reduce your total interest payout on the used car loan.

4. Will the insurance company insure this car and what will the premiums be? Some cars carry premium penalties based on accident and repair records that could run up to 30 percent while other cars carry a special discount based on low repair and accident records and you could save up to 30 percent on premiums. Use this information to your advantage. You can only afford $5000, for example, and the monthly payments of $150 for 36 months put you over that amount. But by buying the Spritzel the insurance agent says you earn a 15 percent discount and the savings over what you now pay for insurance is $15 a month. The savings gives you the money you need to buy the car. Do it.

5. When the lender says the financing rate is 12 percent is that APR? APR stands for annualized percentage rate and when the salesman says you will pay 12

percent make sure you have it in writing that it's 12 percent APR and not 12 percent a month and 120 percent a year. It doesn't hurt when quoted 12 percent a month and payments of $138 a month, to verify with a lending institution that 12 percent translates into $138 a month and $5000 over 36 months. I've had complaints from people who had a salesman tell them the total cost of the car with interest on the loan would be $5000 over 36 months and then tell the customer his or her monthly payments would be $200. Quick arithmetic shows that $200 for 36 months is $7200, not $5000.

6. What, if any, options must you have on the car? When buying a Chevy Caprice in July, a Midwesterner will probably thoroughly check out the air conditioner and forget to look for a rear window defroster.

7. What is the reputation of the car? Check out the car's track record ahead of time. The shiny and clean '82 Cavalier looks cute. But the underpowered engine will be a nightmare.

8. What is the reputation of the dealer selling the car? This is at least as important as the track record of the car. Ask friends, neighbors, co-workers if they've dealt with the dealership before and how they were treated. Call your local Better Business Bureau for a report on the dealer's complaint record. Don't go so much by the number of complaints as by the number resolved versus the number outstanding. How long has the dealer been in business? The longer the better because he probably wouldn't have lasted long pulling lots of tricks.

One word of warning. Some dealers are pillars of the community and are involved in a variety of charitable functions and even go to church on Sundays. But they would negotiate their brother out of $1.50 if given the chance. If the dealer is a member of every local club he's probably never at the dealership. Have a problem and need to see the dealer? You have to join one of his clubs to do so.

Keep in mind, too, the salesmen at the dealership usually are a reflection of the owner. If Sam Slick tackles you as you open the car door when you arrive at the dealership, you can expect a hard sell. Unfortunately, customers sometimes are impressed by the "Salesman of the Month" award or plaque. The award was not bestowed on Sam Slick for being personable. He got it for selling the most cars.

9. Who was the previous owner of the car you're looking at? Some dealers will provide you with the information and you can call the owner. If a potential buyer were to call me I seriously doubt that I'd tell him I didn't change the oil in two years. But the fact the dealer makes the owner available is a positive sign. Some steadfastly refuse to make known the former owner. That would bother me. Keep in mind that the previous owner's name is on the title that you could ask to see.

10. Is the title clear and free from a lien or other problem? Was the vehicle owned by a private party who traded it in or was the prior owner the local rental outfit, police department or cab company? Stay away from buying any car that gets three years use in one. And be cautious of the title when buying from a private party. If Joe Jones is selling the car, the title should be in Joe Jones' name—and there should be a title. No title, no sale. And if the title is

in Joe and Sally Jones name, you watch as Joe and Sally both sign.

11. Why is a one year old Cadillac on the used car lot of a Ford dealer? There could be a variety of reasons and nearly all of them are bad. The owner bought a Cadillac and found he couldn't afford the payments so he went for a cheaper Ford. OK, but if he couldn't afford to own it, could he afford to maintain and service it? Perhaps there was a major mechanical problem and the Cadillac dealer wouldn't take care of it so the owner got angry and jumped over to Ford. Or perhaps the dealer got the car because it was a repo.

12. Why is a one year old Cadillac on the used car lot of the Cadillac dealer? Same as above. Perhaps, however, the owner does trade in on a new car every year. That hardly ever happens anymore, but if the dealer provides you with the previous owner's name you could find out. This is a good time to ask the dealer to run the VIN number through the CRIS system (if it's a one year old Mark VII at a Lincoln-Mercury dealer, use OASIS) to determine if it was a major recall item that prompted the former owner to dump it.

13. If buying from a private party, is there a VIN number on the dash plate and if there is, is the plate free from any marks that indicate it may have been altered? Remember, without the VIN number you could be purchasing a stolen car. If so, you not only will lose the car, you'll lose all the money you paid for it.

14. Is the dealer putting any gimmicky add-ons on the bill of sale? You pay for the car, tax, title and plate

transfer. That's it. No documentary charges. No delivery and handling fee. No used car prep fee. No VHF (Vehicle Handling Fee). You don't want a service contract? Don't accept one. You don't want rustproofing? Don't get it. Since the undercarriage will be covered with grime and grit and the inside of body panels have some gunk or moisture on them, applying rustproofing now makes even less sense than when the car was new and it made no sense then.

15. Was the car a one-owner vehicle? The more owners, the greater the disparity in treatment, service and maintenance. If the car is five years old and already has had three owners you'll have cause for wondering why it was disposed of so many times.

16. What, if any, service was performed on the car before it was put on the lot? Is it the dealer's policy to simply wash and wax the car or send it to a detailer for a quick inside and out cleaning up? Or is there a mechanical inspection performed and what were the results on this car?

17. The used car label sticker on the window says my mechanic can check the car out. Can he take it back to his shop to do it or put it up on the dealer's rack or is he limited to a test drive and visual on-the-ground inspection? The freer the salesman is in allowing an inspection, the better off you are. "You can eye ball it, that's it," may be a warning that any closer inspection would uncover a major problem. Maybe not, but it sure leaves that impression, doesn't it?

18. If there is a warranty on the car, as stated on that window sticker, what is it and what do you pay? If

you really want to make a fool of yourself, let yourself fall victim to the old 50/50 warranty coverage. It sounds good that the dealer will pick up 50 percent of the parts and labor if anything goes wrong. What happens is that if the new brake system costs $200, you'll be charged $400. The dealer pays $200, you pay $200, and unless my arithmetic is off, you really paid 100 percent, not 50 percent.

19. What is meant by credit accident insurance? Sometimes, often when dealerships are dealing with foreign-speaking or low-income buyers, they'll bring a credit accident insurance feature into the contract. What this means is that you take out an insurance policy for X number of dollars a month. In the event you are injured at work and are unable to pay for the car, the insurance pays the dealer the balance due. You are ensuring that no matter what happens to you, the dealer gets paid. They want *you* to pay a monthly premium for this. Think that over twice.

20. What is meant by credit life insurance? Sometimes when foreign-speaking or low-income people are shopping, the salesman will tell them they must buy a credit life policy and pay so many dollars a month for one. In the event you die, the dealer gets his money. Frankly, if I die, I don't give a damn if he gets his money. Let him take the car back, I won't need it any more.

21. Is all the paperwork complete, no items left blank, and do you get a signed copy of all papers before you leave? I am totally amazed at people, from ditch diggers to Ph.D.'s, who will sign a bill of sale or a contract before reading it and before all the empty spaces are either filled in or X'd over so a number can't be put in later at the salesman's whim.

Once all the papers are placed before you, read them *before* you sign. If the salesman hurries you, tell him you either are given uninterrupted time to read the papers now or you take them home and read them at your convenience and will come back, but that you don't sign and certainly don't hand over any money until you've read everything. If he balks, get out. Until he has your check he'll probably back off.

If you come across an item you aren't certain about, get the question answered and have the answer put in writing. No verbal promises.

If the dealer doesn't answer the question to your satisfaction, call your lawyer, the Better Business Bureau, Federal Trade Commission, Attorney General's Office, State Department of Insurance, or anyone who can give you the proper explanation. Most states have dealer trade associations. In Chicago, for example, there's the Chicago Automobile Trade Association that responds to consumer questions and tries to mediate problems with its 700 members.

When you have read and understood all the paperwork, when you have made certain that all blanks are filled, and when the deal is clear in your mind, sign the papers. Keep a copy for yourself. Now make out the check.

Happy motoring.

DIGGING UP THE DIRT

From past accidents to fixing dents

Since I first wrote the original "Used Cars: Finding the Best Buy," some dramatic changes have taken place in the industry to benefit consumers.

One is a means of finding out the history of the used car you have your eye on, including who all the previous owners were and what the odometer reading was each time it changed hands. The other change is a means of eliminating those nagging and nasty little dents and dings in the car body that are such eyesores you either choose not to buy a perfectly good used machine because of it, or you have a hard time finding a buyer for your used car because of that eyesore.

First comes a handy service called Carfax, which is now available not only to car dealers to help them avoid getting ripped off on the car they take in trade, but which is also offered to consumers to keep them from any unexpected and dollar-diminishing surprises.

You have your eye on a used vehicle, but you have to wonder—has it been involved in an accident and fixed with a salvage job held together with glue that's really worth hundreds less than what the seller wants, or is in such bad shape you should simply walk away and look elsewhere?

Or, does the odometer on that car read 35,000 miles, but you suspect the seller could be fudging—by 35,000 additional miles?

With Carfax you can find the answer to these questions

by getting a vehicle history of the car you want to buy before handing over a check.

By calling 1–800–274–2277 between the hours of 8 a.m. and 5 p.m. weekdays you can learn the illustrious or sordid past of the car offered for sale either by a private party or by a dealer on his lot.

You call the number, give the person answering the phone the 17-digit vehicle identification number (VIN), plus the car's make, model, and year. If data on that car is found in the system (Carfax boasts of data on more than 6 million vehicles from 1981 and newer), you are charged a nominal fee on your credit card for the service. If no data is found, there is no charge.

The data lists each owner from the original buyer to present plus the odometer reading at each sale along the line. It also tells you if there is a lien on the vehicle and if the title has been laundered through a variety of stations to clean off the "salvage" designation following an accident. Carfax, however, doesn't provide recall information (see chapter 5 on how to get recall data).

I learned about Carfax in a novel way. One of my daughters was involved in an accident. Her car was totaled, but thanks to the air bag she walked away with only a slight burn on her face from making contact with the nylon cushion.

A few months after the wreck I got a call one evening from the automaker's customer service department in Detroit wanting to know if I was pleased with the service I had gotten on my car in Iowa. I told the young man that 1) I had not visited Iowa in years, and 2) that I couldn't have had the car serviced because a couple of months earlier my daughter had totaled it in a wreck and last I knew it was resting in a junkyard awaiting to be torn apart for parts. But, as long as he called, I asked him to verify the VIN number on the car brought in for service in Iowa with the VIN number on my daughter's car, which I happened to have at hand because I was going through the insurance settlement papers at the time.

The VIN he gave me was the same as my daughter's car, which had been registered in my name and is why I got the call asking if I was pleased with the service.

I was bothered by the call. The car was totaled. I saw the aftermath of what 45 m.p.h. into a mini van making a turn without signaling did to the car. The bag had deployed, meaning the bag, air cannister, and the sensors that activate both were no longer usable. If someone had taken the car and pieced it whole again, chances are there was no usable bag, or cannister or sensors.

I told my story to a car dealer, who said, "Call Carfax," and I did, to find in a matter of minutes via their computer that the vehicle was sold by the junkyard to an out-of-state salvage company. The company obviously pieced it together and sold it to a private party, who perhaps purchased it because it had such low mileage on it, roughly 3,000 over two years' usage. That car should never had been on the road again.

Now I know, and all my friends know, and all my regular column readers know, that before purchasing your next used car, give Carfax a call to learn where it has been before it got to you.

The clock says 35,000 miles yet the computer printout says it had 40,000 miles on it two owners ago? The seller says it's a one owner car and the little old lady only drove it to bingo every Wednesday evening, yet the computer says the last owner was Walt's Speed Shop?

Handy service and well worth the investment, Carfax can save you hundreds if not thousands of dollars in heartache and grief. By the way, the above tip saved you more than the price of this book. Feel a little guilty that you didn't buy two copies and send one to a friend or to the kids. Never too late, friend.

The other bit of advice passed along is not to give up on that used car you like so well just because some careless nitwit swung open his door in the parking lot and left a pockmark on your body panel.

There are a lot of shops which specialize in repairing

those blemishes without drilling a hole in the panel, pulling the dent out, filling the hole in with body filler, and repainting, which, no matter how expert the body man, will still leave a visible scar—especially with today's metallic paints and clearcoat finishes.

The outfit we are most familiar with is Dentbusters, and again we have a daughter to thank for putting us in contact with the firm. This time, however, it was Twin No. 2 daughter who backed out of the driveway and into the recycling bin at the end of the blacktop. Nice dimple in the back end of a Blazer from that adventure.

The Blazer was red and the repair and painting would leave a mark. Told a dealer of the woe and he said, "Call Dentbusters. They specialize in paintless repair. It's where I take my trade-ins to have those pockmarks removed from doors—especially a Mercedes or Beemer in which most folks won't tolerate a ding on a $30,000 used car."

We did and found that what Dentbusters does is take long metal rods or prongs to remove the dent. The prongs resemble those tools the dentist uses to poke around your teeth.

Unlike the dentist's prongs, the Dentbuster ones are about three feet long. The worker slips a prong through one of the many openings in the car door, such as through the opening for the window, and with a little pressure, a little twisting and turning, a little manipulating and massaging the sheetmetal from the inside of the panel, what had been a dent is once again smooth.

No drilling, no hole, no body filler, no touch-up painting that serves as a walking billboard to others that body work has been performed.

You must have patience in waiting for the car to be done because the worker plying the prongs to remove the dent does so very slowly. But I'm talking patience in terms of maybe 30 minutes to an hour for a door ding and you can then drive the car away, whereas at a normal body shop you probably have to leave the vehicle at least a day and then wait for the paint to dry on a ding that has been drilled and repainted — and leaves a scar when done.

As I said, I'm familiar with Dentbusters in the Chicago suburbs. Check the Yellow Pages in your community under car repairs or body shops for the same or similar service. Or, if you know a car dealer or salesman, ask for the name of a service such as this where you live.

Before leaving your car, stop in to see some examples of their work first. You can call your local Better Business Bureau, too, to find the complaint record of the shops in your area.

Keep such a service in mind, too, in buying the used car. If the only problem with the vehicle you want is a couple of dents or dings, you simply tell the seller, "Yes, I like the car, but it will cost me money to have those dents removed, so I'm only going to offer you . . . for the car."

As a rule, paintless repair shops such as Dentbusters are less expensive than body shops, so you can negotiate a lower price on the vehicle you want to buy, take some of that money and have the repair made, and pocket the savings.

One last bit of advice: Never let my daughters drive your car.

Selling Scams

When to buy and when to run

The only difference between a lovelorn columnist and an automotive columnist is that the lovelorn columnist tries to settle disputes among husbands and wives; the automotive columnist tries to resolve problems with Chevys and Fords — and Chryslers and Toyotas and . . . you get the picture. I get hundreds of letters and phone calls each week from consumers with problems. What puzzles me is that so many of the complaints are similar, only the name of the car or the dealership or the mechanic is different.

And while nothing is more frustrating than to come across an obvious scam or con or unethical if not illegal practice, it is equally unsettling that in some cases the consumer *asked* to be put in a position where he or she is going to be or already has been ripped off.

A wise sage once said that the nice thing about having someone stand there and continuously hit you over the head with a 2 × 4 is that it feels so good when they stop.

If consumers would stop and look and see the 2 × 4 coming, maybe they wouldn't suffer so much in the first place.

So here's my list of the TOP 10 signals or clues that someone is about to tattoo your skull with the lumber, obvious warnings that your well being is not utmost on the mind of the sales staff trying to unload a used (and the warnings apply to new vehicles as well) car on you.

1. When the salesman says, "What will it take for you to buy a car today?" tell him it will take a price quote and no more hedging and stumbling waiting for you to make the first move so he knows how much you have or want to spend.

He's asking that question to find out how much dough you are willing to part with because once he finds out, he's gonna get every last penny and probably some more. If you tell him you're interested in what he has to offer for $5,000, how do you know he's not going to show you every car he has on the lot valued at $4,000, but get $5,000 out of you. It would be better to find out first what kind of vehicle $5,000 will buy and then move from there.

2. When a salesman says you can't test drive a car without making a deposit first or that he won't quote you a price until you place a deposit, turn on your heels and leave—without making a deposit.

He wants you to put money down without having a clue if the vehicle is any good because even if the car isn't any good he's going to simply move you through the inventory until you cave in and buy something he has. And what he has in addition to a lot full of used vehicles is your money. Want it back? Have to buy a car to do so.

When you walk into Sears or WalMart does a clerk at the door take your wallet or purse? When you visit a restaurant do you hand the cashier your money and then wait in line for a table? The time to hand over money is when you get something in return. By giving the salesman a deposit you've committed to buying from him—anything.

3. If the salesman says your deposit is refundable even though the sales contract says it isn't, have him scratch out the sentence that says the deposit isn't refundable and write in "refundable in full." Then

you, the salesman, and the owner of the store should initial the change.

Only make a deposit for an amount you are willing to lose if you back out of the deal. Say you see a used Camaro convertible and put down $1,000 and are told to come back tomorrow to pick it up once it's been cleaned up. You go home and have second thoughts, especially after you learn from your friend who also has a Camaro convertible that the cost of insurance is more than the price of the car.

If the salesman says, "It's our policy that you put down $1,000" or that you put down 10 or 20 percent of the price, tell him it's your policy to put down $25 or $50.

4. A salesman might say, "Let me show you what we have in stock, but, oh, almost forgot, to save time later why don't you sign this sales contract now."

You can't believe how many consumers will reach for a pen and ink a contract and they haven't even kicked a tire or slammed a door as yet. Save time? You wouldn't be carting the family from dealership to dealership on a beautiful sunny afternoon if your purpose was to conserve time.

Again, you signed a paper and now all the salesman has to do is fill in the name of a car and the price you owe and he's made a sale. Save time? Sure, his time.

5. Insist on a price quote and not "that '88 Taurus LX will cost you only $200 a month." If he says the Taurus will only cost you $200 a month, the attraction is almost too good to resist because the payments would be so low, unless you hit the calculator keys to learn that $200 a month for 48 months is $9,600—and the car typically retails on any lot for $5,500.

If you only have $5,000 or $6,000 or $7,000 to spend, you want to know which vehicles are available at that price and you'll never know if the only amount the salesman tells you is the cost per month.

6. When the salesman says that $5,500 Taurus will cost you $200 a month, reach for the calculator. Don't enter a showroom or used car lot without one. With the calculator you will be able to punch a few keys and then show the salesman that while he WANTS $5,500 for the Taurus, if you pay him $200 a month for 48 months he will GET $9,600, and since your brain is still functioning, you aren't going to pay that amount.

7. Drive the vehicle you're interested in buying. Okay, it's a red Mustang or Camaro convertible and your friends will simply drool when they see you behind the wheel on a warm July evening, but if the engine is on its last legs, the transmission running on borrowed time, and the body underneath deep into the throes of cancerous rust, the only time your friends will see you behind the wheel is when you go out and sit in it in the driveway because you don't dare or perhaps simply can't move it out onto the roadway.

An unbelievable number of people buy a car without ever turning the key on. It looks good, and the salesman sounded so sincere and honest, how could anything be wrong with it. They buy now, cry later. Drive it, just like you are going to drive it if you buy it. City traffic, expressway traffic, punch the accelerator, kick the brake pedal. Better to find out the problems before you hand over a check.

If there are engine/trans/brakes/body problems, walk away. If the engine/brakes/trans/body are in good shape but there is a minor problem — poor radio reception, buttons missing, window catches and won't roll down, broken head or taillight, outside mirror missing — have it fixed NOW, not after you hand over the money.

8. Remember there is no three-day cooling off period on the purchase of a used car (or new one either). The cooling off period only applies when someone comes to the door of your home or apartment to solicit a sale. Probably 25 percent of the calls I get every Monday morning are from folks who bought a car on Saturday and now want to return it on Monday.

9. If you get a flier in the mail advising you that you are one of a select group of customers chosen to take advantage of a special one or two day used vehicle sale, immediately take that flier and deposit it in the garbage.

The dealer purchases a mailing list, often based on car registrations, and the reason you got one is that you are one of 10,000 people within a 20-mile radius of that store who owns a five-year-old car, making you a primary candidate to be in the market for another vehicle after that length of time.

The dealer is counting on volume to make money, a bunch of rubes showing up at the door with flier in hand thinking they are going to get the best of it because of a special clearance. He might get 100 people in the door without a flier, 500 with a flier. That's why the flier.

And when the flier says you are going to get one of five prizes listed — a year-long vacation in Florida, a condo on the beach in Hawaii, a 185-foot-long ship complete with crew, a 97-inch color TV and free cable the rest of your life, or a five-piece screwdriver set, which of those five prizes do you actually think you are going to take home with you?

If you aren't sure of the answer, please stop reading now, grab the phone, and make an appointment with a shrink, because you're in greater need of medical advice than you are car buying advice.

Some fliers will lead consumers to believe that the salesman doesn't know a sale is going on and striking a deal, you can hand him that piece of paper you got in the mail that looks like a check and you'll get an extra $100 discount. Wow, will the saleman be surprised and shocked when you hand him that wannabe check. Sure, the owner of the store spent $2,500 in a mailing to promote a sale and he never told his sales staff about it. Be serious, folks.

10. Sign nothing until after you've struck a deal with the salesman and he's detailed each and every charge that you owe—price of the car including the rate of financing and the total financing cost over the life of the contract, tax, title, license fee, and warranty cost if any.

After the sale is not the time to be informed that the finance rate is 24 percent for 60 months when you were told it was 12 percent for 48 months, or that there is a $40 documentary fee for filling out your paperwork, or there is a prep fee for cleaning the car before you turn on the key.

THE most frequent complaint from consumers is problems with the paperwork. Most sign the documents without reading, without noticing there are blank spots to be filled in (by the salesman), without checking that the $5,500 total price with tax, title, and license fee that he just quoted you is in the space and not $6,500 instead.

The typical caller will say, "A couple of days after I got the car home I opened the glove box and looked at the papers and he charged me $1,000 more than he said he would," or "He said the loan would be for 36 months, but. . . ." or "He said the interest rate would be. . . ." or "He said he was throwing in the warranty for free, but. . . ." or "He said he was allowing me $2,000 for my trade-in, but the paper says he only gave me. . . ."

Why sign a paper that you didn't read for an amount you can't afford and then claim you were taken advantage of?

Avoid these 10 pitfalls and you'll save some money — and lots of grief. And for being so nice in agreeing to heed the warning signs of a ripoff, you get a special bonus. Check out the next chapter.

Repair Ripoffs

Okay, here's your bonus, some tips on how to avoid another ripoff once you take that used car you've just bought in for service.

When it comes to service:

1. When you pull into the shop and the mechanic or service writer says, "all cars like that do that," often what he is trying to tell you is that he hasn't the foggiest notion of what is wrong, but rather than let you in on his stupidity, he's gonna blame the manufacturer for producing cars "that do that."

Flee the shop.

2. Remember that it's not unusual for parts to wear and components to break regardless of mileage. Countless people call or write me whining that, "Can you believe the fuel pump broke and I only have 95,000 miles on the car?"

It's not up to the factory to replace each and every worn or broken part and if you write a letter demanding that Ford or Chevy or Toyota give you a new fuel pump after "only 95,000 miles of driving" then not only are you going to be out a fuel pump, you're going to be out a 32-cent stamp.

3. When the mechanic says the squeak or rattle is normal, move on to another service shop. Either the person doesn't know what he's doing or he knows but doesn't want to make the repair because it involves too much time and labor that he could be devoting to "quicky" repairs that will earn him more money in the long run.

4. When you drive into a service shop and tell the person in charge to, "do what it takes to make it better again," don't cry and complain when the good doctor of mechanics hands you a bill three hours later for $1,000. Give a person carte blanche to pull and replace and he's going to take advantage of it — and you.

5. When you pull into a transmission, brake, engine repair shop or dealership and the person on duty tells you that a new transmission/brakes/engine are needed before a service worker even lifts the hood or drives the car, leave at once or be prepared to be handed a bill equal to the national debt for work that probably didn't need to be done in the first place. If your mechanic is a psychic, he should have his own 900-number and be doing an infomercial on cable TV rather than repairing cars.

6. When you pull into the transmission/brake/engine repair shop or dealership and the person on duty says it will cost you $1,000 to fix your vehicle before he's lifted the hood, pulled a wheel, or driven the car, get out of there fast because the only thing he's after is your money. How does he know the cost when he doesn't know what's wrong?

7. Don't pull in for service on a Monday morning, tell the mechanic the car "makes a noise" and expect to have the car repaired and be on your way home before you finish a cup of coffee in the waiting room. Be prepared

to tell the mechanic what the noise sounds like (clunk, rattle, squeak) and when it happens (starting the car, when braking, when making a turn), as well as adding any other specifics (only at 45 miles per hour and only from the left side of the front end).

8. Don't expect your vehicle to remain trouble-free or last very long if you only change oil based on seasons. Change oil and filter every 3,000 miles or three months—whichever comes first. Only driving your car a few times or a few miles a month doesn't prolong the life, in most cases it actually shortens it from such things as condensation and or deposit buildup from never operating your car at proper temperature.

9. A defect is when the factory designs or builds the vehicle incorrectly. Neglect is when the owner doesn't treat the car correctly and avoids periodic service and maintenance.

10. When you visit your doctor for a check-up and the guy waiting ahead of you in line has heart trouble, does that mean you have heart trouble, too?

Of course not, so why when you pull into a service shop and the guy ahead of you with the same make of car has a problem with his whizofram do you then insist you have a problem with your whizofram, too, and then suspect all cars of the same make have whizofram trouble?

You could have pulled into the shop for an oil change, but the guy ahead of you says, "Brought my car in 'cause the whizofram is actin' up. You got the same model, bet your whizofram is actin' up, too."

If your whizofram was working fine, it probably isn't now, or at least you've been convinced it doesn't based on the tale of woe from the other guy in the shop.

Right after neglect, the power of suggestion creates more claims of car ailments than actual defects.

INDEX
